3rd International Conference on Advanced Materials Research and Manufacturing Technologies (AMRMT 2018)

IOP Conference Series: Materials Science and Engineering Volume 436

Shanghai, China
11 – 13 August 2018

ISBN: 978-1-5108-7654-5
ISSN: 1757-8981

Printed from e-media with permission by:

Curran Associates, Inc.
57 Morehouse Lane
Red Hook, NY 12571

Some format issues inherent in the e-media version may also appear in this print version.

Printed by Curran Associates, Inc. (2019)

For permission requests, please contact the Institute of Physics at the address below.

Institute of Physics
Dirac House, Temple Back
Bristol BS1 6BE UK

Phone: 44 1 17 929 7481
Fax: 44 1 17 920 0979

techtracking@iop.org

Additional copies of this publication are available from:

Curran Associates, Inc.
57 Morehouse Lane
Red Hook, NY 12571 USA
Phone: 845-758-0400
Fax: 845-758-2633
Email: curran@proceedings.com
Web: www.proceedings.com

TABLE OF CONTENTS

IOP Conf. Series: Materials Science and Engineering **436** (2018) 011001 doi:10.1088/1757-899X/436/1/011001

PREFACE

AMRMT 2018 was held in Shanghai, China during Agu.11-13, 2018. AMRMT 2018 was organized by Hong Kong Society of Mechanical Engineers. The conference provides a useful and wide platform both for display the latest research and for exchange of research results and thoughts in Advanced Materials Research and Manufacturing Technologies. The participants of the conference were from almost every part of the world, with background of either academia or industry, even well-known enterprise. The success and prosperity of the conference is reflected high level of the papers received.

The proceedings are a compilation of the accepted papers and represent an interesting outcome of the conference. This book covers 2 chapters: Materials, Design and Applications; Materials Processing.

We would like to acknowledge all of those who supported AMRMT 2018. Each individual and institutional help were very important for the success of this conference. Especially we would like to thank the organizing committee for their valuable advices in the organization and helpful peer review of the papers.

We sincerely hope that AMRMT 2018 will be a forum for excellent discussions that will put forward new ideas and promote collaborative researches. We are sure that the proceedings will serve as an important research source of references and the knowledge, which will lead to not only scientific and engineering progress but also other new products and processes.

Prof. Alan Kin-Tak Lau, Swinburne University of Technology, Australia

IOP Conf. Series: Materials Science and Engineering **436** (2018) 011002 doi:10.1088/1757-899X/436/1/011002

COMMITTEES

Conference Chair

Prof. Alan Kin-Tak Lau, Swinburne University of Technology, Australia

Program Chair

Prof. Gong Hao, National University of Singapore, Singapore

Advisory Committee

Prof. Rossi Setchi, Cardiff University, UK

Prof. Jennifer Loy, Griffith University, Australia

Prof. Dong Ruan, Swinburne University of Technology, Australia

Prof. Min Wang, The University of Hong Kong, Hong Kong SAR

Prof. Haitao Huang, Hong Kong Polytechnic University, Hong Kong SAR

Prof. Jun Wang, The University of New South Wales, Australia

Prof. Zhengwei You, Donghua University, China

Prof. Jae-Won Choi, The University of Akron, USA

Prof. Pei Li, The Hong Kong Polytechnic University, Hong Kong SAR

Dr. Yu Dong, Curtin University, Australia

Prof. Shaoyun Fu, Chongqing University, China

Prof. Debes Bhattacharyya, University of Auckland, New Zealand

Prof. Joong Hee Lee, Chonbuk National university, Korea

Prof. Ada Pui Ying Hung, Swinburne University of Technology, Australia

Prof. Xiao Feng Li, Shanghai Maritime University, China

Clara Chan, Lee Kee Group, Hong Kong SAR

Sam TS Chin, Hong Kong Aircraft Engineering Company Limited, Hong Kong SAR

Prof. Jinsong Leng, Harbin Institute of Technology, China

Prof. Thanchanok Ratvijitvech, Mahidol University, Thailand

Prof. Nat Vorayos, Chiang Mai University, Thailand

Prof. Watchanida Chinpa, Prince of Songkla University, Thailand

IOP Conf. Series: Materials Science and Engineering **436** (2018) 011002 doi:10.1088/1757-899X/436/1/011002

Prof. Chatchai Putson, Prince of Songkla University, Thailand

Prof. Varunee Premanond, King Mongkut's University of Technology Thonburi, Thailand

Dr. Hing-Ho Tsang, Swinburne University of Technology, Australia

International Technical Committee

Prof. Teik-Thye Lim, Nanyang Technological University, Singapore

Prof. Christopher Cheeseman, Imperial College London, UK

Prof. Zhongwen Yao, Queen's University, Canada

Dr. Mihai Caleap, University of Bristol, UK

Dr. Grzegorz Lisak, Nanyang Technological University, Singapore

Dr. Prof. Qian Shunzhi, Nanyang Technological University, Singapore

Dr. Prabhudev M.S, Governement Polytechnic Kalgi, India

IOP Conf. Series: Materials Science and Engineering **436** (2018) 011003 doi:10.1088/1757-899X/436/1/011003

Peer review statement

All papers published in this volume of *IOP Conference Series: Materials Science and Engineering* have been peer reviewed through processes administered by the proceedings Editors. Reviews were conducted by expert referees to the professional and scientific standards expected of a proceedings journal published by IOP Publishing.

Microstructure and Mechanical Properties of Ultra-High Strength Pearlitic Wire in Single-pass Cold Drawing

M Z Wei[1], Z H Chen[1], X M Zhang[1], D P Wang[1], K Sun[1], J Zhou[1,2,*] and F Xue[1,2]

[1] Jiangsu Key Laboratory for Advanced Metallic Materials, Southeast University, Nanjing 211189, China
[2] Jiangsu Key Laboratory of Advanced Structural Materials and Application Technology, Nanjing Institute of Technology, Nanjing 211189, China

[*] jethro@seu.edu.cn.

Abstract. The effect of compression ratio on the microstructure and mechanical properties of ultra-high strength pearlitic wire in single-pass cold drawing has been studied. The ultra-high tensile strength cord steel with a carbon content of 1.02 wt% were pulled through a single-pass drawing to different compression ratios (0-30%). With the increase of the single-pass compression ratio, the tensile strength, microhardness and torsional properties of the steel wire increased gradually. When the compression rate increased from 0 to 30%, the tensile strength of the steel wire increased from 1263.4 MPa to 1553.7 MPa, the torsional strength of steel wire increased from 1126.83 MPa to 1279.66 MPa. In the single-pass drawing process, the strengthening mechanism of the pearlitic steel wire are mainly the work hardening caused by the dislocation slip in the ferrite, and the fine grain strengthening caused by the decrease of the pearlite sheet spacing.

1. Introduction

Pearlitic steel wire of high-strength is a nano-crystalline structure metal material which is obtained by strong plastic deformation [1]. And now pearlitic steel wires are applied to engineering applications in large-scale. Among them, Development of steel wire for cold-drawing bridge cables and steel cords for radial tires are toward to positive high strength and ultra-high strength [2-3].

Radial tires replace oblique rubber tires was a major revolution in car tires. Radial tires have the advantages of good wear resistance, low rolling resistance, fuel economy, good cushioning performance, good heat dissipation, high speed, high safety and reliability [4]. According to different materials of carcass and belt of tires, radial tires can be divided into three types: all steel radial tires (TBR), semi-steel radial tires (PCR) and all-fiber radial tires [5]. There are many researches on pearlite steel wire containing 0.82% carbon and 0.92% carbon at home and abroad, but fewer studies and reports on the mechanical properties and microstructure of ultra-high tensile strength steel wires in single-pass cold drawing have been carried out. In this paper, the hypereutectoid pearlitic steel wire with carbon content of 1.02wt% was used as the raw material, and the mechanical properties and microstructure evolution of the steel wire with different strains in the single-pass drawing process were studied.

2. Materials and method

The object studied here is an ultra-high tensile strength cord steel with a carbon content of 1.02 wt%. The sample preparation process is mainly to pull Φ5.5mm wire rods through a single-pass drawing to different compression ratios (10%-30%).The chemical composition of 1.02 wt% pearlitic steel wire used in the test is shown in table 1:

Table 1. 1.02 wt% Pearlite Steel Chemical Composition

Element	Content(wt%)
C	1.02
Si	0.215
Mn	0.310
P	0.0089
S	0.0041
Cr	0.297

First, the steel wire was cut into small pieces. After polished and etched, they were observed with a metallurgical microscope. Corroded samples were observed under scanning electron microscopy (SEM) for the microstructure of the pearlite and the pearlite wire spacing was measured also. In this paper, transmission electron microscopy (TEM) was used to observe the change of pearlite lamellae and cementite lamellae and the distribution of dislocations in ferrite with increasing drawing strain. When performing tensile strength test, the gauge length of the tensile sample was selected to be 100mm, and the tensile strain rate was 10mm/min. The experimental parameters of the torsion test in this article were determined according to the national standard and the actual test conditions.

3. Result

3.1 The metallographic microstructures of ultra-high tensile strength cord steel

Because the pearlite grains are very small, the scanning electron microscope with higher resolution was used to observe the change of the microscopic morphology of the steel wire with different compressibilities. The SEM image of initial wire rod is shown in figure 1. From figure 1, it can be seen that there are obvious lamellar structures and pearlite groups, and the lamellar orientation between adjacent groups of pearlite is random.

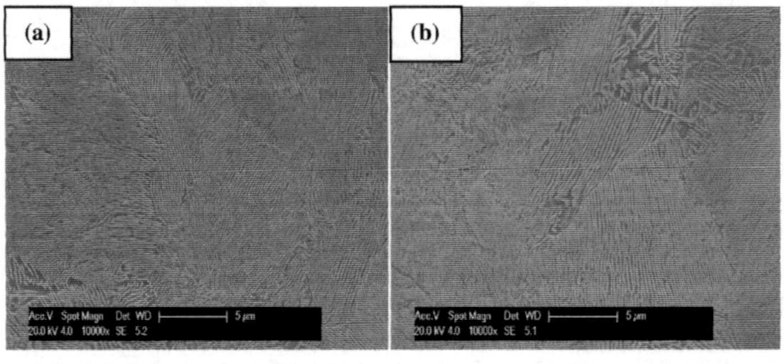

Figure 1. SEM image of initial wire rods: (a) Cross section; (b) Longitudinal section

The SEM images of steel wires with different compression ratio of 10% and 30% in the cross section is shown in figure 2. According to the figure, it can be seen that with the increase of the strain in the single-pass drawing, part of the pearlite sheet has seen bent and folded, as shown by the arrows in the figure. At the same time, the distance between layers of pearlite became smaller.

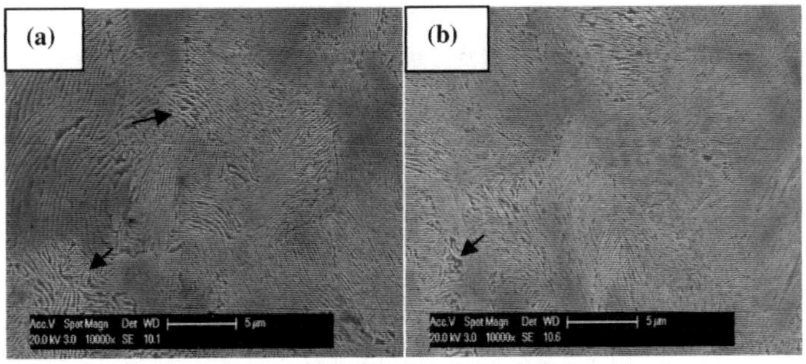

Figure 2. The SEM images of steel wires with different compression rates in the cross section: (a) Compression rate is 10%; (b) Compression rate is 30%

The TEM images of hypereutectoid pearlitic steel wires with a carbon content of 1.02% is shown in figure 3. Obvious pearlite sheets can be seen in the figure, the black area is a cementite sheet and the bright white area is pearlite. Comparing figure 3(a) and figure 3(b), it can be clearly seen that, compared to the initial steel wire, the pearlite sheet spacing decreases and the cementite sheets are thinned as the compression ratio increases. Besides, minor deformation has appeared in the cementite sheet. However, due to the strain is small, the change is not obvious.

The TEM images of the cementite sheets at a compression rate of 30% is shown in figure 3(c). Figure 3(d) is an enlarged view of the red block in figure 3(c). The cementite sheets in the figure 3(d) are clear and intact, with no fragmentation, indicating that cementite did not fragment when the strain was small.

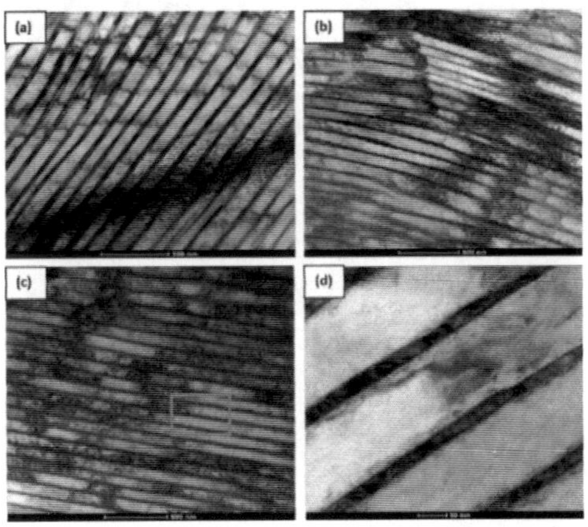

Figure 3. TEM images of steel wires with different compressibility:
(a)The compression rate was 0; (b) The compression rate was 10%; (c) The compression rate was 30%; (d) The cementite strip with a compression rate of 30%

3.2 The mechanical properties of single-pass drawing steel wire

The initial diameter of wire rods used for single-pass drawing is 5.5 mm. The tensile strength of steel wires under different drawing strains are listed in table 2. In order to clearly show the relationship between tensile strength and strain, the data in the table is plotted as a curve as shown in figure 4. From the figure, it can be seen that as the strain for single-pass drawing increases, the tensile strength of the pearlitic steel wire slowly increases first, then increases rapidly, and gradually tends to be flatten finally. When the compression ratio of the steel wire increased from 0 to 30%, the tensile strength of the steel wire increased from 1263.4 MPa to 1553.7 MPa, and the tensile strength increased by 22.98%.

Table 2. Relationship between tensile strength and compressibility of single-pass drawing steel wire

Compression ratio (%)	Strength (MPa)
0	1263.4
10	1267.3
15	1281.5
20	1468
25	1537
30	1553.7

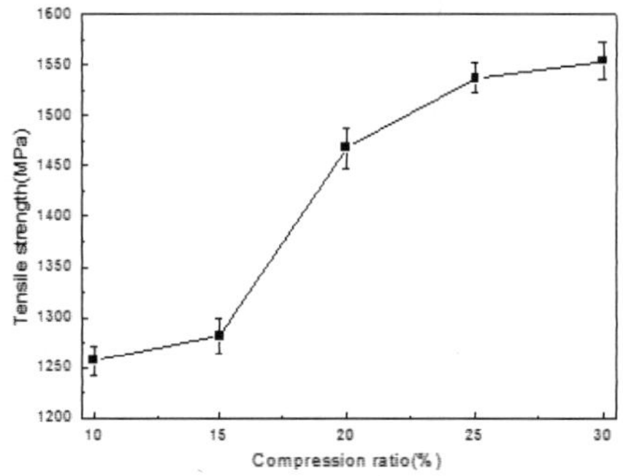

Figure 4. Relationship between tensile strength and compression ratio of Single-pass drawing steel wire

The results of torsion test of the single-pass drawing steel wire are shown in table 3. The number of torsion turns of the original wire rod was 15 and the torsional strength was 1126.83 MPa. From the data in the table, it can be seen that as the single-pass drawing compression rate increases, the number of twisted turns of the steel wire gradually increases, and the torsional strength also increases. When the compression rate increased from 10% to 30%, the number of torsional turns of the steel wire increased from 15 to 21 turns, and the torsional strength increased from 1137.31 MPa to 1279.66 MPa.

Table 3. The torsion test results of single-pass drawing steel wire

Compression ratio(%)	Diameter(mm)	Torsional turns	Torsional strength(MPa)
0	5.5	15	1126.83
10	5.22	15	1137.31
15	5.07	15	1142.94
20	4.92	18	1175.76
25	4.76	18	1176.77
30	4.60	21	1279.66

4. Discussion

4.1 Effect of strain on microstructure in single pass drawing

From the results of SEM and TEM, it can be seen that with the increase of the strain, the interlamellar spacing of the pearlite groups became smaller, the alternating cementite and ferrite layers in the pearlite deformed at the same time. Pearlite is the product of the eutectoid reaction in the Fe-C binary system. The ferrite in the pearlitic steel is a solid solution of C in Fe and belongs to the body-centered cubic structure. It has good plasticity and can achieve plastic deformation at high temperature. However, cementite is a hard and brittle M3C interstitial compound. Traditional physical metallurgy believes that such mesophases cannot be plastically deformed at room temperature. The previous literature reported [6-8] that when the pearlitic steel was cold-worked, only the ferrite was deformed in the microstructure, and the cementite was not deformed. If the stress was high, the cementite layer would cracked and crushed. In recent years, in the literature on the study of the microstructure of large deformation of cold-drawn steel wire [9-11], the observation results of simultaneous deformation of cementite and ferrite have been reported, and this is referred to as "coordinated deformation".

4.2 Effect of strain on mechanical properties in single-pass drawing

Pearlitic steel belongs to high carbon steel and its strength is higher than that of mild steel mainly due to the "second phase" strengthening of cementite. In this paper, the tensile strength, torsional strength and hardness of the pearlitic steel wire increased with the drawing strain after single-pass drawing, but the plasticity decreased a lot. This is similar to the change in mechanical properties of ordinary metal materials during deformation processing. For metallic materials of cubic structure, strengthening after cold working is generally attributed to "work hardening." In this paper, this strengthening process occurs only in ferrite.

The crystal structure of ferrite belongs to the typical body-centered cubic structure. The deformation of the cubic metal is mainly controlled by the slip mechanism. In the single-pass drawing process, when the external force reached the critical resolved shear stress of ferrite, one of the slip system of ferrite began to slip. At this time, ferrite was in a single slip state, dislocations increased a lot and the dislocations slide along a certain direction in a certain sliding plane. Since the dislocations were prone to slide at this time, the resistance is less, so the steel wire strength increased slowly when the compression ratio was small. As the compressibility and the critical shear stress increased, multiple slip systems in ferrite started. At this point, the resistance to dislocation movement increased, making the slipping process more difficult. At the macro level, the plasticity was reduced and the strength was improved. The TEM images of the steel wire with 10% compression rate and 30% are shown in figures 5(a) and (b). Comparing the two figures, it can be seen that the density of dislocations in ferrite increased significantly as the compressibility changed.

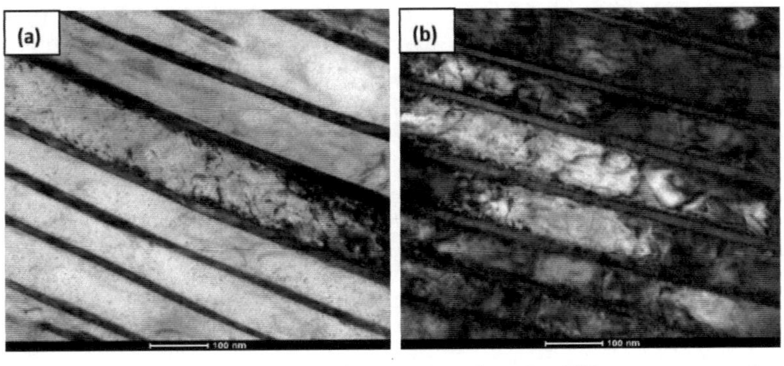

Figure 5. The dislocation changes of 1.02wt% wires for different compression rates

(a) The compression rate is 10%; (b) The compression rate is 30%

In addition to the deformation hardening of ferrite, fine-grain strengthening is also produced during the deformation of cold-drawn pearlite. From the TEM images shown in figure 5, the pearlite lamellar spacing decreased as the strain increased. According to Hall-Petch formula, it can be seen that the smaller the pearlite sheet spacing, the higher the steel wire strength. Therefore, the decrease of the spacing between pearlite layers due to cold drawing deformation is also one of the factors causing steel wire strengthening.

5. Conclusion

With the increases of the single-pass compression ratio, the microstructure of the wire changes. The pearlite bends and folds on the cross-section and the pearlite structure undergoes coordinated deformation on the longitudinal section. The pearlite sheet spacing is slightly reduced, the cementite sheet is thin and slightly deformed, but the cementite sheet remains intact at this time. With the increased of the single-pass compression ratio, the tensile strength and torsional properties of the steel wire increased gradually. When the compression rate increased from 0 to 30%, the tensile strength of the steel wire increased from 1263.4 MPa to 1553.7 MPa, the torsional strength of steel wire increased from 1126.83 MPa to 1279.66 MPa. In the single-pass drawing process, the strengthening mechanism of the pearlitic steel wire are mainly the work hardening caused by the dislocation slip in the ferrite, and the fine grain strengthening caused by the decrease of the pearlite sheet spacing.

Acknowledgement
This study was supported by the National Natural Science Foundation of China (51004039) and the Opening Project of Jiangsu Key Laboratory of Advanced Structural Materials and Application Technology (ASMA201602).

References
[1] Baraz V R, Sokolov A A and Belov A V 1995 *Met. Sci. Heat. Treat* **37(8)** 339-40
[2] Zelin M 2002 *Acta. Mater* **50(17)** 4431-47
[3] Nam W J, Bae C M, Oh S J and Kwon S J 2000 *Scripta. Mater* **42(5)** 457-63
[4] Borchers C, Al-Kassab T, Goto S and Kirchheim R 2009 *Mater. Sci. Eng.* A **502(1–2)** 131-8
[5] Embury J D and Fisher R M 1966 *Acta. Mater* **14(2)** 147-59
[6] Chandhok V K, Kasak A and Hirth J P 1966 *ASM Trans* **59(2)** 288-301
[7] Porter D A, Easterling K E and Smith G D W 1978 *Acta. Mater* **26(9)** 1405-22
[8] Nikolussi M, Leineweber A and Mittemeijer E J 2008 *Acta. Mater* **56(3)** 5837-44
[9] Langford G 1970 *Metall. Mater. Trans* A **1(2)** 465-77
[10] Norimitsu Koga 2012 *Scripta. Mater* **67(4)** 400-3
[11] Puttick K E 1957 *J. Iron. Steel Inst* **185(7)** 161-76

The Effect of Parameter of MIG Arc Brazing on Microstructure and Bonding Strength of Tin-based Babbitt Alloy ZChSnSb11-6 on ASTM 1045 Steel

K Sun[1], P Y Wang[2], M Z Wei[1], J Zhou[1,3,*] and F Xue[1,3]

[1] Jiangsu Key Laboratory for Advanced Metallic Materials, Southeast University, Nanjing 211189, China
[2] Jiangsu Key Laboratory for Advanced Biologic Materials, Nanjing Normal University, Nanjing 211189, China
[3] Jiangsu Key Laboratory of Advanced Structural Materials and Application Technology, Nanjing Institute of Technology, Nanjing 211189, China

*jethro@seu.edu.cn.

Abstract. The effect of parameter of MIG arc brazing on the microstructure and the bonding strength of ZChSnSb11-6 and steel substrate has been studied. A cladding of ZChSnSb11-6 was obtained on the surface of medium carbon steel by MIG arc brazing. The microstructure of ZChSnSb11-6 and the microstructure of the interface between ZChSnSb11-6 and steel substrate have been systematically investigated. The highest bonding strength of ZChSnSb11-6 and the steel substrate is 66.42Mpa when welding voltage is 22V and welding current is 120A. The reason for the enhancement of the bonding strength was discussed. When the hard and brittle SnSb phase near the IMC layer is subjected to tensile stress at the interface between the Tin-based Babbitt alloy and the steel substrate, stress concentration occurs and the initiation and propagation of cracks are promoted. When the welding voltage is 20V, The number of the SnSb phase near the IMC layer decreases while the welding current increases, so the bonding strength increases. When the welding current is 120A, the distribution of SnSb phase changes while the welding voltage increases. When the welding voltage is 22V, the bonding strength is the highest.

1. Introduction

As a support shaft transmission part, the bearing bush is very important, especially in large workpieces, the performance of the bearing bush affects the stability and reliability of the entire transmission system. The bearing bush mainly play the role of anti-friction, that is, to deal with the large impact, friction to ensure the integrity of the bearings [1-2]. Compared with other bearing alloys, Tin-based Babbitt alloys have been widely used in key components of bearings due to their unique excellent corrosion resistance, especially in steam turbines, high-speed engines and compressors [3-4]. Moreover, Tin-based Babbitt alloys even can remain soft after repeated cycles of loading because of its lower melting point [5-7].

At present, Tin-based Babbitt alloys on steel-base bearings are mainly produced by centrifugal casting [8]. However, Tin-based Babbitt alloys is easily separated due to the centrifugal force generated during the centrifugal casting process, resulting in non-uniform microstructure distribution and unstable performance [9]. In this paper, MIG welding method was selected to bond Tin-based Babbitt alloys to steel substrate. The effects of different welding parameters on the microstructure and morphology of the interface IMC were investigated.

2. Materials and Method

In this paper, Babbitt welding wires of ZChSnSb11-6 (named as BW11-6) was used. The chemical composition of Sn-based Babbitt Alloys ZChSnSb11-6 was shown in Table 1. The diameter of the welding wire was 1.6 mm. The ASTM 1045 steel was used as the steel substrate. Before welding test, the surface of substrate was polished to remove the oxidation layer. The surface of welding wires were cleaning to remove oil.

Table 1. Chemical composition of Sn-based Babbitt Alloys ZChSnSb11-6 (wt%).

Brand	Sn	Pb	Cu	Zn	Al	Sb	Fe	Bi	As	others
BW11-6	Bal.	0.35	5.5-6.5	0.01	0.01	10.0-12.0	0.1	0.03	0.1	0.55

There were 5 sets of welding parameters for testing, numbers of samples and corresponding parameters were shown in Table 2. Protective gas was pure argon and welding speed was 40cm/min. The specimens were cut into squares, the specimens were polished in advance and then corroded by 4% nitric acid alcohol solution. The microstructures of the Babbitt and the interface were studied by the optical microscopy (OM) and scanning electron microscopy (SEM).

Table 2. Welding parameter of Sn-based Babbitt Alloys ZChSnSb11-6 (wt%).

Sample	Welding voltage (V)	Welding current(A)
1	20	115
2	20	120
3	20	125
4	22	120
5	24	120

In this experiment, the bonding strength of Babbitt alloy and ASTM1045 steel was tested according to the destructive testing method of the bimetallic bonding strength of sliding bearing which referred to the national standard GB/T12948-1991. The digital picture of sample was shown in Figure 1. The tensile speed was controlled by means of stress, and the applied value of stress was 10 N/(mm²·s). Increase the stress until the alloy layer or the bond layer broke, or the alloy layer peeled off from the steel back. Five replicate tests were performed on each group of samples. The relative error of the value was less than 10% .

Figure 1. digital picture of the bonding strength test specimen

3. Results

3.1 Microstructures results of Sn-based Babbitt alloys

The metallographic structure of the Sn-based Babbitt alloys BW11-6 studied in this paper was shown

IOP Conf. Series: Materials Science and Engineering **436** (2018) 012002 doi:10.1088/1757-899X/436/1/012002

in Figure 2(a)(b)(c)(d)(e). It can be seen from Figure 2 that BW11-6 is composed of a large number of square SnSb phases, acicular or star-shaped Cu6Sn5 phases in the α-Sn phases.

The microstructure of the Sn-based Babbitt alloys BW11-6 obtained from the five kinds of welding parameters was shown in Figure 2. When the welding voltage is 20V, the metallurgical structure is very sensitive to the changes in the welding current, with welding current increased from 115A to 125A, the crystal grains become coarser, especially the blocky SnSb phase. When the welding current is 120A, the grain size is relatively small when the welding voltage is 22V. At the same time, the microstructure is the most uniform at 22V.

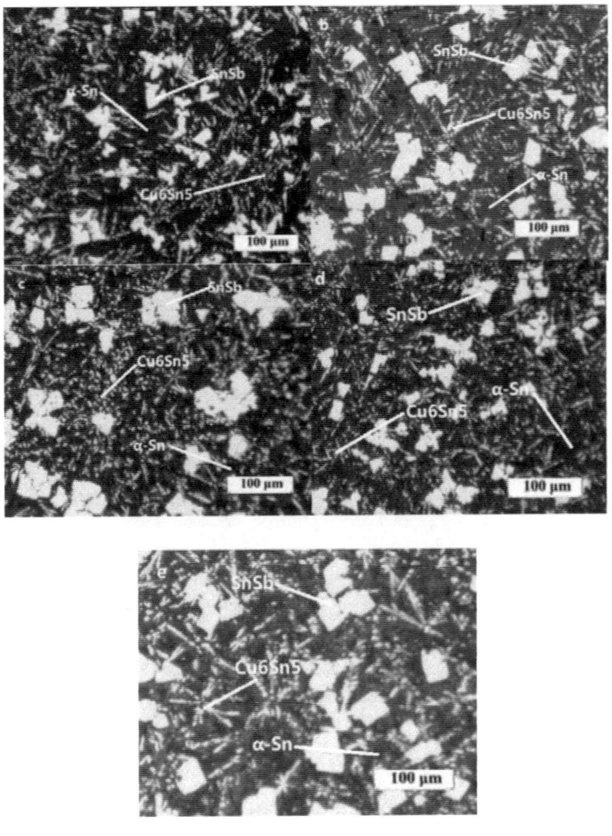

Figure 2. Microstructures of different welding parameter of Sn-based Babbitt alloys BW11-6:(a) sample 1(b) sample 2(c) sample 3(d) sample 4(e) sample 5

3.2 Interfacial structure of the Sn-based Babbitt/ASTM 1045 steel interface

The bonding strength of BW11-6 and the steel substrate depends on the microstructure of the interface. The interface microstructure of the five welded joints is shown in Figure 3(a)(b)(c)(d)(e). In Figure 3, there is an obvious transition layer between the substrate and the Babbitt alloy, and a large amount of bulk compound (SnSb) is accumulated near the transition layer.

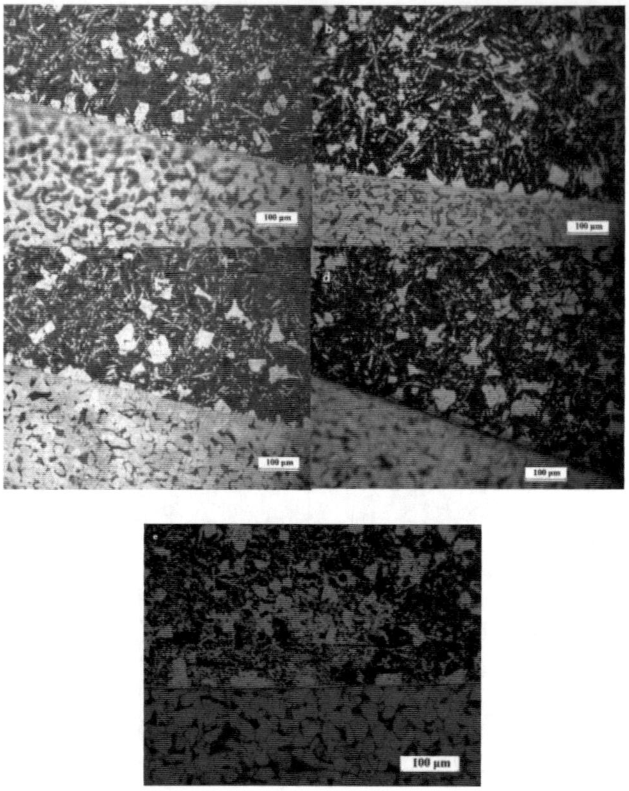

Figure 3. Interfacial structure of different experiment conditions: (a) sample 1 (b) sample 2 (c) sample 3 (d) sample 4 (e) sample 5

SEM and EDS were used to further analyze the interfacial microstructure and composition of the samples. It was found that the SEM images of the interfacial microstructure were similar, so one of the samples was selected for the study. The SEM image of the interface of is shown in the left column of Figure 4(a) and the right column of Figure 4 is the result of the line scanning. The results of EDS of points 1, 2 and 3 is shown in Table 3.According to Figure 4(b), the IMC layer at the interface of BW11-6 with the steel substrate is very thin (about 2 microns).

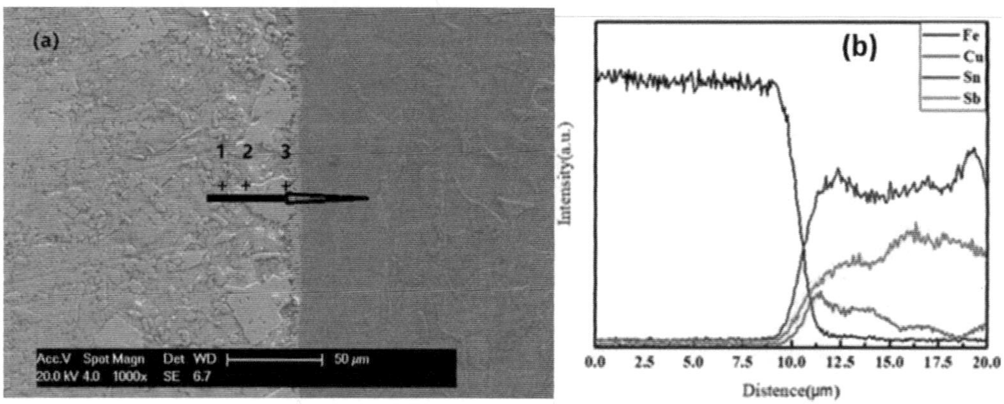

Figure 4. The SEM image of interfaces of Sn-based Babbitt alloys and steel substrate with EDS analysis and line scanning results

Table 3. The results of EDS at different positions in the IMC layer (at%).

Position	Fe	Sn	Cu	Sb
1	1.34	78.75	1.34	18.57
2	3.05	47.72	45.59	3.63
3	22.30	38.78	33.56	5.36

3.3 The mechanical properties of MIG arc brazing Sn-based Babbitt

The average test results of the bonding strengths of different samples in the test are shown in Figure 5. When welding voltage is 22V and welding current is 120A, the bonding strength of BW11-6 is the highest, which is 66.42MPa. According to the results of sample 1, 2 and 3, it can be seen that when the welding voltage is 20V, the bonding strength increases from 41.33 Mpa to 62 .74 Mpa as the welding current increases from 115 A to 125 A. According to the results of sample 2, 4, and 5. It can be seen that when the welding current is 120A, as the welding voltage increases from 20V to 24V, the bonding strength first increases from 58.35 Mpa to 66.42 Mpa and then decreases to 60.71 Mpa.

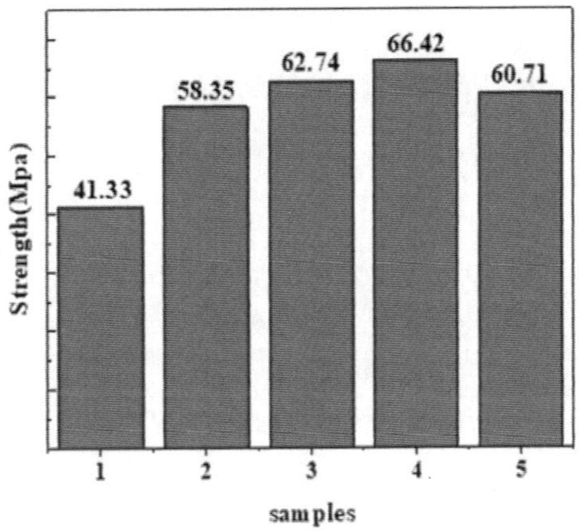

Figure 5. The bonding strength of Sn-based Babbitt and steel substrate

4. Discussion

4.1 The effect of different welding parameter on microstructure of Sn-based Babbitt alloy

From Figure 2, when the welding voltage is 20V, as the welding current increases, the square hard phase SnSb grain is larger. At the interface between BW11-6 and the steel substrate, the phase SnSb floats easily due to the low density and causes gravity segregation. The relatively large proportion of needle-like Cu6Sn5 itself has the ability to prevent the SnSb from floating, so even though the SnSb grains become larger, the larger Cu6Sn5 can better block the SnSb from floating, resulting in a reduction in the number of SnSb at the interface between BW11-6 and the steel substrate. The change of the welding voltage will affect the stability of the welding process. This affects the homogeneity of the microstructure of the Babbitt alloy. When the welding voltage is 22V, the crystal grains is the most uniform, and the distribution of SnSb phase at the interface between BW11-6 and the steel substrate is also the most uniform.

4.2 The effect of different welding parameter on mechanical properties of Sn-based Babbitt alloy

When welding current increases, the number of SnSb phases at the interface between BW11-6 and the steel substrate decreases. When the hard SnSb phase is subjected to tensile stress, stress concentration occurs. So the aggregation and growth of a large number of SnSb phases at the interface between BW11-6 and the steel substrate promote the initiation and propagation of cracks and reduce the bonding strength.

The welding voltage mainly influences the stability of the welding process. Inhomogeneous microstructures results in the aggregation of SnSb phase at certain junctions, which reduce the bond strength. When welding current is 120 A, the welding voltage is 22V, the bonding strength of BW11-6 and steel substrate is the highest. This is because the microstructure of the joint is the most uniform when the welding voltage is 22V.

5. Conclusion

Under different welding parameters, Tin-based Babbitt alloys BW11-6 has different microstructures, and the bonding strength between steel substrate is also different under MIG welding. When the welding voltage is 22V and the welding current is 120A, the bonding strength is the highest. It is 66.42Mpa.SnSb phase near the IMC layer reduces the bonding strength of the sample in the interface between the Tin-based Babbitt alloys and the steel substrate. Due to the existence of SnSb phase, when the specimen is subjected to tensile stress, stress concentration occurs, and the initiation and propagation of cracks are promoted.

Acknowledgement

This study was supported by the National Natural Science Foundation of China (51004039) and the Opening Project of Jiangsu Key Laboratory of Advanced Structural Materials and Application Technology (ASMA201602).

References

[1] Goudarzi M M, Jahromi S A J and Nazarboland A 2009 Mater. Des **30** 2283-88
[2] Zeren A, Feyzullahoglu E and Zeren M 2007 Mater. Des **28** 318-23
[3] Valeeva A K, Valeev I S and Fazlyakhmetov R F 2014 J. Frict. Wear **35** 311-5
[4] Sadykov F A, Barykin N P and Valeev I S 2003 J. Mater. Eng. Perform **12** 29-36
[5] Khonsari M M, Booser E R 2002 J. Tribol **124** 428
[6] Rigney D A 1997 Tribol. Int **30** 361-7
[7] Bo J 2003 Tribol. Int **36** 781-9
[8] He Y, Zhao Z and Luo T 2013 Mater. Des **52** 923-31
[9] Kamal M, El-Bediwi A and Lashin A R 2011 Mater. Sci. Eng. A **530(24)** 327-32

Microscopic and Macroscopic Bipolar Injection and Carrier Recombination in Single-layer Si Nanocrystals

Jie Xu[1,*], **Shulin Zheng**[1], **Yuxing Chen**[1], **Wei Li**[2] **and Jun Xu**[2]

[1] College of Electronic and Optical Engineering, College of Microelectronics, Nanjing University of Posts and Telecommunications, Nanjing 210023, China
[2] National Laboratory of Solid State Microstructures, School of Electronic Science and Engineering, Nanjing University, Nanjing 210093, China

*jiexu@njupt.edu.cn

Abstract. Si nanocrystals (NCs) based optoelectronic devices have attracted more and more attentions due to the quasi-direct energy band structure and the adjustable band gap of Si NCs. In these optoelectronic devices, the carrier injection and recombination process is a crucial issue and needs to be deeply understood. Here, bipolar carrier injection and recombination process in a single layer of Si NCs was observed at the microscopic and macroscopic scales, by electrostatic force microscopy (EFM) and alternating-current electroluminescence (ac EL) measurement, respectively. EFM result showed that opposite charges could be injected into the single-layer Si NCs and they would laterally recombine with each other. On the other hand, the single-layer Si NCs also showed a frequency dependent EL intensity, which was attributed to the time limitation of carrier injection and radiative recombination process.

1. Introduction

Bulk Si has an indirect band gap structure that strongly limits its optoelectronic application. However, thanks to the quantum confinement effect, a quasi-direct band structure and size dependent band gap can be expected in Si nanocrystals (NCs), which facilitates the preparation and application of Si NCs-based light emitting diodes (LEDs) and other optoelectronic devices [1-4]. In these devices, carrier injection and recombination process determines the device performance and needs to be deeply studied and understood [5]. In recent years, electrostatic force microscopy (EFM) and Kelvin-probe force microscopy (KFM) have been extensively introduced to visualize the charge injection and spread process in Si NCs based materials, giving detailed information at the nanoscale [6-8]. More recently, we have demonstrated the quantitative relationship between the EFM/KFM signal and sample surface potential, enabling quantitative analysis of surface charge distribution from the microscopy signals [9, 10]. On the other hand, alternating-current electroluminescence (ac EL) detection has been proposed as a powerful tool to observe the radiative recombination process at the macroscopic scale, and even found helpful to enhance the optoelectronic device performance [11, 12]. In the present work, the carrier transportation process of single-layer Si NCs was observed by EFM and ac EL measurement, from which the bipolar injection and recombination behaviors were clearly shown at the microscopic and macroscopic scales respectively.

2. Experiments

2.1. Sample Preparation

Single-layer Si NCs sandwiched by amorphous Si carbide (a-SiC) layers were prepared by excimer laser crystallization technique [13]. First, a-SiC/a-Si/a-SiC sandwiched structures were deposited on p-Si (100) substrate (1.5–3 Ω-cm) by a conventional plasma enhanced chemical vapor deposition (PECVD) system. Silane (SiH_4) was used for a-Si growth while methane (CH_4) and SiH_4 ($[CH_4]:[SiH_4]=10:1$) gases for a-SiC growth. The thickness of bottom a-SiC and a-Si layers was 10 nm and 4 nm, while that of top a-Si layer was 2 nm (sample A) or 35 nm (sample B). After PECVD growth, both samples were exposed under a KrF excimer laser (Coherent, COMPexPro 201F, 248 nm) to crystallize the a-Si layer and form Si NCs. Finally, plane aluminum electrode was evaporated on the backside of substrate of both samples, and transparent circular indium-tin-oxide (ITO) electrode with a diameter of 1.5 mm was sputtered on the top of sample B. The sample structures are shown in figure 1. In the following experiment, sample A was used for the EFM measurement, and sample B for the ac EL detection.

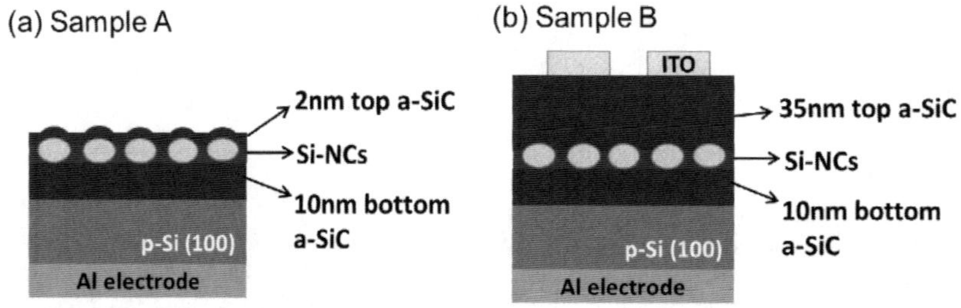

Figure 1. The sample structure of (a) sample A and (b) sample B.

2.2. EFM Measurement

EFM measurement (Bruker, Nanoscope IIID multimode) was then carried out for sample A to visualize the carrier injection and recombination process. Excess carriers were first injected into Si NCs from the biased conductive tip (Bruker, SCM-PIT type) in a modified tapping mode, where "TM Deflection" parameter was used for feedback control instead of "Amplitude Setpoint" [5], as shown in figure 2(a). The tip scanning rate is 1 Hz per line during the charge injection operation. After charge injection, EFM measurement was implemented in a two-pass interleave scan. In the first pass, the tip scanned the sample surface to obtain topography lines; in the second pass, the tip was lifted 50 nm above the sample surface under a constant bias of +4 V, and finally the tip resonant frequency shift was output as EFM signal, as shown in figure 2(b).

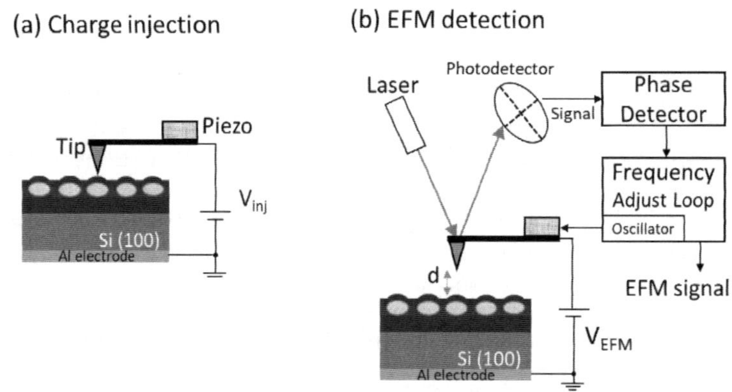

Figure 2. (a) The microscopic charge injection operation through a biased tip. (b) EFM measurement in non-contact mode.

2.3. AC EL Measurement

Sample B was driven by a sine ac voltage source (Vpp=14V) that was generated by a source signal generator (Agilent 33210A) connecting to a signal amplifier (Agilent 33502A). The ac EL signal was detected by a CCD detector (HORIBA Jobin Yvon) in the visible range. The EL spectra was calibrated considering the sensitivity of the detector system.

3. Results and discussion

3.1. Sample structure

Figure 3 shows the atomic force microscopy (AFM) images of sample A before and after excimer laser irradiation. It is shown that the as deposited sample has a smooth surface (rms roughness smaller than 0.2 nm), while after laser annealing, grain-like Si NCs form and they make the surface textured (rms roughness about 1 nm).

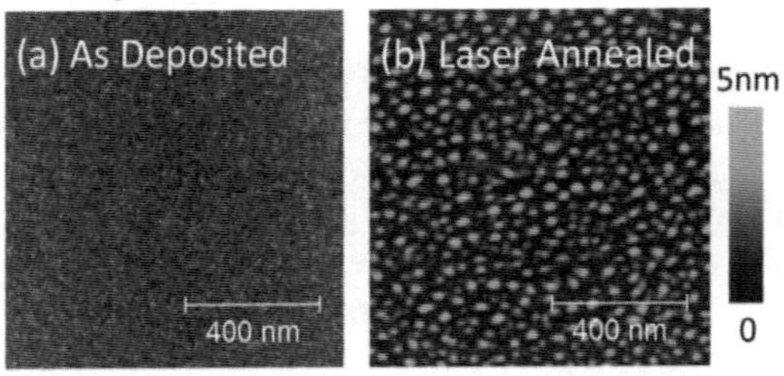

Figure 3. AFM topography of sample A (a) before and (b) after excimer laser irradiation.

Moreover, Raman spectrum also proves the crystallization of a-Si layer (not shown here). The as deposited sample only shows an amorphous band around 480 cm^{-1}, however, after laser annealing a nanocrystalline peak at 515 cm^{-1} appears which confirms the formation of Si NCs [14].

3.2. EFM Results

Figure 4 shows the EFM results after bipolar injection through the biased tip. In figure 4(a), a square area of 1.5×1.5 μm^2 was first charged under −3 V tip bias, and then the inner square area of 0.3×0.3 μm^2 was recharged under +3 V tip bias. It is shown that the EFM signal Δf reduces in the negatively charged area while it increases in the positive one. Considering that the EFM voltage V_{EFM} is positive (+4 V) in the measurement, and that Δf is proportional to the product of V_{EFM} and the excess charge quantity [6, 10], it can be deduced that electrons were injected into Si NCs during the −3 V injection procedure, and opposite charges (holes) were injected during the following +3 V injection process.

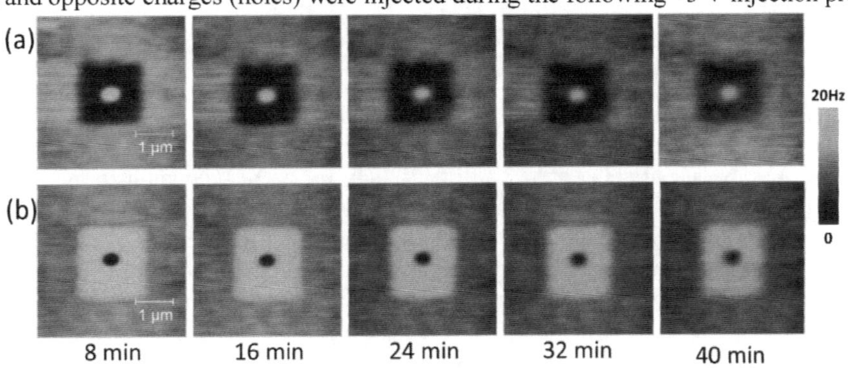

Figure 4. EFM images of sample A after bipolar injection. (a) First −3 V injection in the 1.5×1.5 μm^2 area and then +3 V injection in the inside 0.3×0.3 μm^2 area. (b) Reversed bias polarity compared to (a). V_{EFM} equals to +4 V in all the images.

It should be noticed here that although the injection bias intensity (3 V) and nominal tip scanning rate (1 Hz per line) are the same, the latter recharging area is much smaller and thus the actual charging scan rate per length is only one fifth of the former charging process, making it possible for electron extraction and hole injection. Figure 4(b) shows a similar result when the injection bias polarity is reversed. The EFM images clearly show the bipolar injection into Si NCs and the opposite-type carriers form a core-ring shape distribution.

The profiles of figure 4 at different elapsed times are plotted in figure 5(a) and (b). It is found that the EFM signal intensity of charged regions gradually decreases to the uncharged level, indicating that there exists charge leakage in the Si NCs. A possible charge leakage channel is through the bottom a-SiC layer to the substrate, which was pointed out by our previous work [5]. Besides, the signal decay is much faster in the charged core area than the surrounding ring area, in both figure 5(a) and (b). The central carriers (holes in figure 5(a) or electrons in figure 5(b)) almost disappear after 40 min, however, the surrounding electrons in figure 5(a) or holes in figure 5(b) still remain about one half after the same period. This phenomenon indicates that in addition to vertical charge leakage through the bottom a-SiC layer to substrate, lateral charge movement and recombination occurred after bipolar injection into Si NCs, shown in figure 5(c). First, the surrounding carriers move to the core due to Coulomb attractive interaction; then, opposite carriers recombine with each other which accelerates the EFM signal decay of the core area. The recombination mechanism may be radiative, non-radiative or even Auger recombination, and the EL detection below shows that radiative recombination is possible in the single-layer Si NCs.

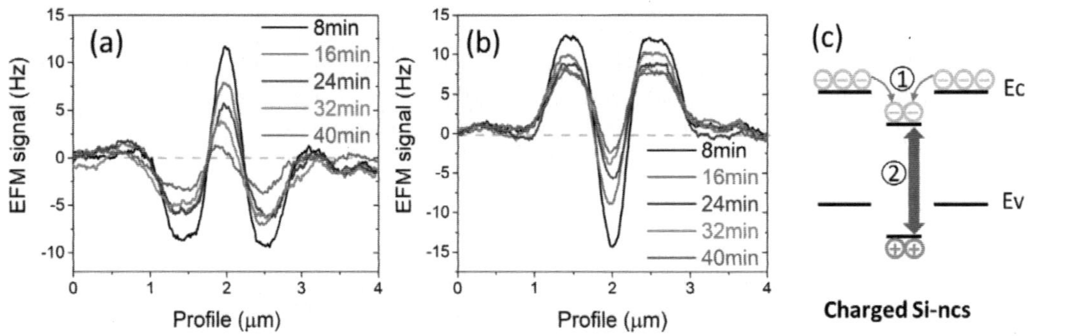

Figure 5. (a) The cross-sectional profiles of figure 4(a) at different times. (b) The profiles of figure 4(b). (c) The energy band diagram of Si NCs in figure 4(a), and the carrier movement includes ① the lateral transportation and ② carrier recombination.

3.3 AC EL results

EFM measurement shows at the microscopic scale that both electrons and holes can be injected into Si NCs simultaneously, and the carriers will move towards and recombine with each other even when they stay at separate NCs at first. In order to further study the carrier recombination mechanism, EL detection is carried out and it is interesting to discuss the EL spectra under ac drive voltages.

The ac EL spectra of sample B driven by ac voltage sources at different frequencies are shown in figure 6(a). The ac EL spectra shape is very similar to the dc EL spectrum of Si NCs/a-SiC system [14], indicating that they have similar bipolar carrier injection and recombination mechanism. The spectra has a wide band in the visible range, which possibly comes from the radiative recombination of electron-hole pairs in the Si NCs. The photon energy at the peak wavelength (730 nm) is 1.70 eV, which is close to the band-to-band energy for Si NCs embedded in SiC matrix with a diameter of 4 nm [14], and the certain size distribution makes the luminescence peak expanded to a wide band. It is also worth noting here that the EL spectra could only obtained in the forward bias condition (positive bias applied to the p-Si substrate), implying that holes and electrons are injected from the substrate and

ITO electrode respectively. On the other hand, the infrared emission may come from the radiative recombination from Si NCs/a-SiC interface state or band tail states of a-SiC [5].

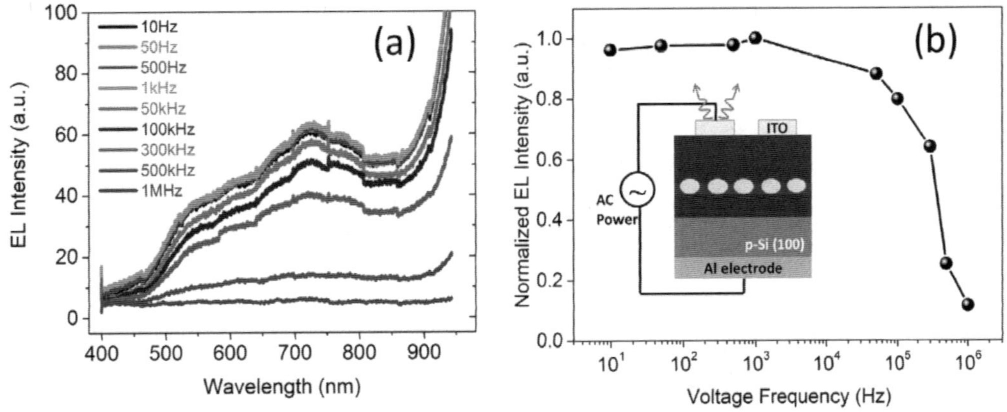

Figure 6. (a) The ac EL spectra of sample B at different drive frequencies. The peak to peak voltage V_{pp} is 14 V. (b) Normalized ac EL intensity at different drive frequencies. The inset is the detection configuration.

Figure 6(b) shows that the EL intensity is dependent on the drive frequency. The EL intensity and shape barely change when the drive frequency increases from 10 Hz to 1 kHz, however, the EL intensity gradually decreases when the drive frequency reaches 50 kHz and it totally disappears at 1 MHz. The recombination mechanism is thought unchanged at different drive frequencies, but under high frequency voltage conditions the carriers do not have enough time to tunnel through the a-SiC layer into Si NCs. Besides, the radiative recombination lifetime of Si NCs is about several microseconds [15], and thus the bias at high drive frequency will also extract the injected carriers and drop their recombination chance. Therefore, it can be deduced that the time needed for carriers injection and recombination process of our sample is about 20 μs (1/50 kHz), which is consistent with the time-resolved photoluminescence (TRPL) measurements of Si NCs [16].

Figure 7 shows the supposed energy band diagram of the EL emission process. First, electrons and holes were injected from ITO electrode and substrate, respectively, through Fowler-Nordheim tunneling under a large forward bias. Second, electrons and holes relaxed to the band edge from the high energy state. Finally, electron-hole pairs radiatively recombined with each other and emitted a light band centered at 730 nm. Besides, the infrared light emission might derived from the carrier recombination from interface or band tail states.

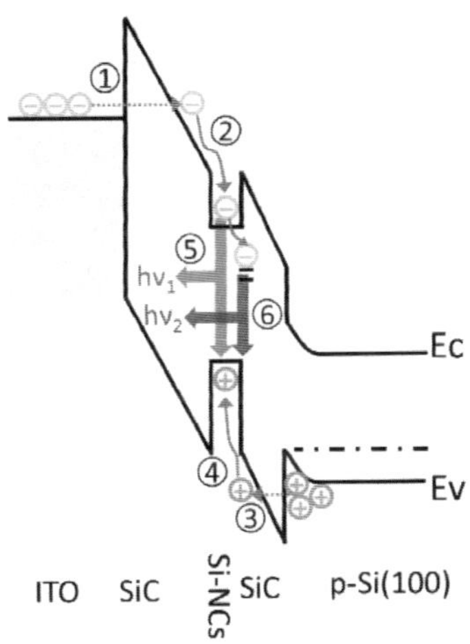

Figure 7. The energy band diagram of the bipolar carrier injection and recombination in the single-layer Si NCs. ① ③ : carrier tunneling through the a-SiC layer; ② ④ : carrier relaxation process; ⑤: visible light emission from band-to-band recombination in Si NCs; ⑥: infrared light emission from the interface or band tail states.

4. Conclusions

Single-layer Si NCs were prepared by laser crystallization technique. The bipolar injection and carrier recombination process was observed by EFM and ac EL measurement at the microscopic and macroscopic scales, respectively. EFM result shows that electrons and holes can be simultaneously injected into Si NCs by a biased tip in contact with the sample surface. Adjacent opposite charges will attract and recombine with each other due to Coulomb interaction, and it accelerates the charge leakage process. To further study the carrier recombination mechanism, ac EL detection was also carried out. It is shown that electrons and holes will be injected from top electrode and substrate respectively, and finally recombine to emit visible light. The EL intensity gradually diminishes when the drive frequency increases to 50 kHz, which implies a time limitation (about 20 μs) needed for the carrier injection and recombination process.

Acknowledgements

This work is supported by Natural Science Foundation of Jiangsu Province (No. BK20160909), NUPTSF (Nos. NY215039 and NY217041) and STITP 2018 of NUPT (No. 399). The authors also want to thank Dr. Dongke Li, Dr. Zewen Lin and Dr. Jingjing Liu for their helpful discussions.

References

[1] Maier-Flaig F, Rinck J, Stephan M, Bocksrocker T, Bruns M, Kubel C, Powell A K, Ozin G A and Lemmer U 2013 *Nano Lett.* **13** 475-80
[2] Priolo F, Gregorkiewicz T, Galli M and Krauss T F 2014 *Nat. Nanotechnol.* **9** 19-32
[3] Yanagawa H, Inoue A, Sugimoto H, Shioi M and Fujii M 2017 *J. Appl. Phys.* **122** 223101
[4] Zhang P, Li D K, Jiang L Y and Xu J 2017 *J. Phys.: Conf. Ser.* **844** 012012
[5] Xu J, Ji Y, Lu P, Bai G, Ren Q and Xu J 2018 *AIP Adv.* **8** 015224
[6] Mélin T, Diesinger H, Deresmes D and Stiévenard D 2004 *Phys. Rev. Lett.* **92** 166101
[7] Borowik Ł, Kusiaku K, Deresmes D, Théron D, Diesinger H, Mélin T, Nguyen-Tran T and Roca i Cabarrocas P 2010 *Phys. Rev. B* **82** 073302
[8] Melitz W, Shen J, Kummel A C and Lee S 2011 *Surf. Sci. Rep.* **66** 1-27
[9] Xu J, Wu Y, Li W and Xu J 2017 *Nanotechnology* **28** 365705
[10] Xu J, Chen D, Li W and Xu J 2018 *J. Appl. Phys.* **123** 184301
[11] Mu W, Zhang P, Xu J, Sun S, Xu J, Li W and Chen K 2014 *IEEE J. Sel. Top. Quant.* **20** 206-11
[12] Liu J, Sheng X, Wu Y, Li D, Bao J, Ji Y, Lin Z, Xu X, Yu L, Xu J and Chen K 2018 *Adv. Opt. Mater.* **6** 1700897
[13] Xu J, Xu J, Wang Y, Cao Y, Li W, Yu L and Chen K 2014 *Nanotechnology* **25** 055703
[14] Rui Y, Li S, Xu J, Cao Y, Li W and Chen K 2012 *J. Non-Cryst. Solids* **358** 2114-7
[15] Walters R J, Bourianoff G I and Atwater H A 2005 *Nat. Mater.* **4** 143-6
[16] Lu P, Li D, Zhang P, Tan D, Mu W, Xu J, Li W and Chen K 2016 *Opt. Mater. Express* **6** 3233

IOP Conf. Series: Materials Science and Engineering **436** (2018) 012004 doi:10.1088/1757-899X/436/1/012004

Cell Behaviour of the Biomimetic Modifications on a New TiHfNb Alloy Developed for Orthopaedic Applications

Virginia Paredes[1,2,3,*], Emiliano Salvagni[2] and José María Manero[3]

[1]Department of Mechanical Engineering, Universidad del Norte, km 5 Vía Pto Colombia, Barranquilla, Colombia
[2] Institute of Advanced Chemistry of Catalonia (IQAC-CSIC), Barcelona, Spain
[3]Nanoengineering Research Centre (CRnE), Barcelona

*Email: paredesv@uninorte.edu.co

Abstract. The field of cellular behaviour and the biomimetic modifications on inorganic materials destined for biomedical applications shows a remarkable increase in scientific studies and, therefore, this is reflected in the investment priorities and the fundamental effort of the global research community in this area. The essential purpose of this study is to assessment the cellular behaviour of a novel TiHfNb alloy with biomimetic modification. For the mimic surface, the samples were initially bioactive with oxygen plasma or with piranha, then the following samples were biofunctionalized with APTES + maleimide and finally different surface peptide sequences were immobilized (RGD, FHRRIKA, PHSRN or 50/50 mixtures of RGD / FHRRIKA or RGD / PHSRN). To evaluate the biomimetic modification, adhesion and cell proliferation studies were carried out. Different techniques for chemical characterization were used, including: X-ray photoelectron spectroscopy (XPS), contact angle (CA), immunofluorescence, scanning electron microscope (SEM) and lactate dehydrogenase (LDH). With respect to the different peptide sequences used, the results indicated that the samples with RGD and the mixture exhibited more extended cells on the surface.

1. Introduction

Currently, there is an increase in the number of implants required by the population worldwide. In addition, patients who need these implants desire to maintain the same level of activity and quality of life. The implants have been manufactured with different materials, such as metals and their alloys, polymers, ceramics and composite materials. Then, a biomaterial design as an implant must provide mechanical stability, and durable bonding with the surrounding bone in a process called osseointegration. In addition, the biomaterial should avoid adverse tissue reactions and minimize the absence of infection. The most common biomaterials are metals based on Ti, Co and steel, because they have excellent mechanical properties. The polymers are the second group, the last group is composed of ceramic materials and less frequent compounds [1,2]

The success of an implant is conditioned to its acceptance by living tissues; as well as the formation of new bone around the implant. Several studies have reported that the bone-implant interface depends on the biological factors and factors associated with the patient. Therefore, it is important to know the design of the implant and its surface, the distribution of loads between the bone and the implant and the surgical procedure. Then, it is necessary to investigate the different strategies that modify the surface of the implant to determine if through these modifications it is possible to obtain the best cellular response.[3] There are different methods to obtain modifications in the properties of the surface of the material; these promotes the cellular response, therefore, the validation of the biocompatibility of new surfaces could be based on experimental cellular studies on biomimetic surfaces.[3–5] Among the biomaterials, Ti and its alloys have shown a good combination between biocompatibility, corrosion resistance and lower density and lower modulus of elasticity.[2] For this reason, the aim of this study is to analyze the cellular response in a nobel Ti alloy (TiHfNb) with a low elastic modulus. Its surface was biomodified through the immobilization of different short sequences of peptides; it was carried out through the process of surface activation, silanization and immobilization of biomolecules. The variables that defined the behavior of the cells were number of adhered cells and their cell morphology, which are indicators of adaptation to the environment. Therefore, the biological behaviors on short peptide sequences were used to validate the biomimetic modification.

2. Materials and Methods.

2.1 Materials

Ti alloy studied (Ti-16Hf-24Nb) was fabricated into UPC Laboratory. The material was cutted into 2 mm of thickness. Each sample was polished using 600 and 1,200 grit silicon carbide (SiC), abrasive papers and finally, they were further polished with napped polishing pads with a colloidal alumina polishing abrasive (1 and 0.05 µm). Disks were degreased from macroscopic contamination by ultrasonication in: acetone, water and ethanol, and then dried with compressed air. Before the activation process, a chemical cleaning was made by ultrasonication in cyclohexane, isopropanol, ethanol, deionized water and acetone.

2.2 Biomimetic modification

The biomimetic modification involved three step: 1) *Surface activation*: through oxygen plasma (OP),.[6–8] 2) *Silanization*: this stage consisted of incubation for 1h in a pentane basic solution containing the organosilane (3-aminopropyltriethoxysilane (APTES)). Then is modified with 3-(maleimido)-propionic acid N-hydroxysuccinimide ester for crosslinker,[8–10], and 3) *Peptide immobilization*: The samples were incubated overnight (ph = 7), room temperature in a phosphate buffer solution containing the desired peptide (CGGRGDS, CGGPHSRN, CGGFHRRIKA, and two mixtures (50/50) of CGGRGDS/CGGFHRRIKA and CGGRGDS/CGGPHSRN). GenScript provided the peptides used for the study. The strategy was to use biomolecules with cysteine (Cys) and glycine (Gly) as spacers. Because the Cys contains a sulfur atom in its side chain, which is highly reactive. Likewise, the peptides were acetylated (blocked by reacting the terminal amino group), thus, they only have the ability to react with silanes through the side chain of cysteine (-SH).[4,11–14] To the surface characterization, the techniques used were Contact Angle (CA), interferometry and X-ray Photoelectron Spectroscopy (XPS).

2.3 *In vitro biological response:*

All the samples were immersed for 30 min in a phosphate buffer solution (PBS / Invitrogen) and bovine serum albumin (BSA (1%)) in order to avoid unspecific protein absorption. The samples were sterilized in ethanol for 10 min [15], finally the surface were washed three times with sterile PBS.

2.3.1 Cell Adhesion: This process was evaluated to 6h at 37°C, The criteria used were the number of cell and their morphology. Were used these types of samples: 1) Two control: Sample, both samples without treatment (TisB),, and one of them with PBS-BSA (1%), another sterilized with ethanol but without PBS-BSA (1%) (TiB), 2) Samples silanized with peptide sequences are showed at table 1. Samples were placed in each sample 6x103 cells, and it was incubated in serum-free medium. The characterization techniques used were: Immunofluorescence, because the nuclei were stained with DAPI (4,6-diamino-2-phenol-dihydrochloride), the adhered cells were observed in a confocal microscope, which allowed their quantification per unit area. Finally, the SEM images allowed to evaluate cell morphology and determine their area. The nomenclature used Control (Ti), Ti+OP (TiOP), TiOP+APTES+maleimide (TiAM), TiAM+RGD (TiAMR), TiAM+ FHRRIKA (TiAMF), TiAM+PHSRN (TiAMP) TiAM+RGD+FHRRIKA (TiAMRF), TiAM+RGD+PHSRN (TiAMRP)

3. Results and Discussion.

3.1 Biomimetic Modification

Activation Process: The results indicated that Ti have a roughness 26.3 nm an TiOP 22.9nm; Thus, the samples can be considered smooth or minimally rough. [16] The hydrophilicity is associated with the reduction of the contact angle from 62.1° to 5.4°, these results indicated the introduction of hydrophilic groups (OH^-,O^2, H_2O) on the surface.. Therefore, the treated samples seemed to provide excellent results in terms of cleaning and chemical modification of the surface.[6–8] However, the CA technique did not provide information on chemical composition; Then X-ray photoelectron (XPS) spectroscopy was used to evaluate the active functional groups (eg, OH-), since a higher concentration of OH- groups on the surface will benefit the next step of modification process (silanization).[8–10] By XPS was possible to observe as activation method produced an increase of O 1s and also OH- groups; due Ti had 46% of oxygen while TiOP had 58%. The XPS high resolution of O 1s showed three contribution O^{2-}, OH- and H_2O,[17,18], where the percentage of O^{2-} (Ti = 28%, TiOP = 25%) and H_2O (Ti = 8%, TiOP = 5%) were lower after the activation, however, the activation produced an increase of OH^- group ranging from 10% up 28%. The percentage of O 1s together with the OH^-/O^{2-} ratios demonstrated that OP was efficient, due to the OH^-/O^{2-} ratio for TiOP was 1.12 while for the Ti was 0.36.

3.1.1 Silanization Process: After the silanization process, the samples increased the contact angle from 5.4 ° to 75.23 °. These results confirm that the silanization process produced a superficial modification. ByXPS it was possible to confirm the presence of silanes on the surface through the silanized samples showed the presence of Si 2p from 1% for TiOP up to 12% for TiAM. Also there was an intensification of the carbon content from 10% up 55%, due to the carbon chains of the silanes.

3.1.2 Peptide Immobilization: The aim of peptide immobilization on biomaterials is to induce a specific cell adhesion, thereby to improve the tissue-implant interface. The technique used to chemical characterization was XPS; The results are shown in table 1, these allowed confirming the presence of biomolecules because: in all the samples with peptides there was presence of S 2p, also the amount of Si 2p on the surface decreased due it was coated with biomolecules. Finally, there was an increase in the percentage of O 1s. The carboxyl groups (COOH) belong to the terminal chain of the peptide sequences or to the peptide bonds (-OC-NH-). [8,19]

Table 1. Results of Peptide immobilization by XPS (atomic percentage)

Element	TiAM	TiAMR	TiAMF	TiAMP	TiAMRF	TiAMRP
S 2p	0±0.0	1±0.1	1±0.1	1±0.2	1±0.0	1±0.2
Si 2p	12±1.0	10±0.8	8±0.6	8±0.7	10±0.4	8±0.7
O 1s	23±1.9	36±3.2	38±2.9	36±3.3	39±2.7	36±3.1

3.2 In vitro biological response

3.2.1 Cell Adhesion; the staining of the nuclei was carried out by immunofluorescence; through the quantification of the nuclei it was possible to determine the number of cells per cm^2 (figure 1), the best results are obtained in samples containing RGD, either alone or in mixtures.[11,13,20–24] The control samples contain a number of cells similar to that of biofunctionalized samples could be due the Ti alloy surface contains TiO_2 and several studies have reported that this oxide favors the cell adhesion process.[16] Also, when it is evaluated the cellular response, a relationship between the number of adhered cells and their morphology must be established; thereby, SEM images were obtained to visualize the shape of the cells and estimate their area. Figure 2.a shows that the immobilization of biomolecules influences the morphology of the cells. The control sample had a contracted cell and smaller extension area. Then, it is possible affirms that the RGD sequence promotes a more comfortable adhesion, because in surfaces with this sequence (RGDS alone or in mixtures) more widespread cells are observed, which have a larger area, and these exhibit the presence of filopodia. Corroborates what it is expressed by Petershans.et al., who expressed that the RGDS sequence, although it is the smallest motif in the extracellular matrix of fibronectin, acts as the receptor integrin of the cellular link, initiating the formation of local adhesion contacts. The quantification of the cells area i shown in figure 2.b, the best results are shown on surfaces with RGD and mixtures, although the standard deviation has a wide range, and then there are not statistically significant differences between them.

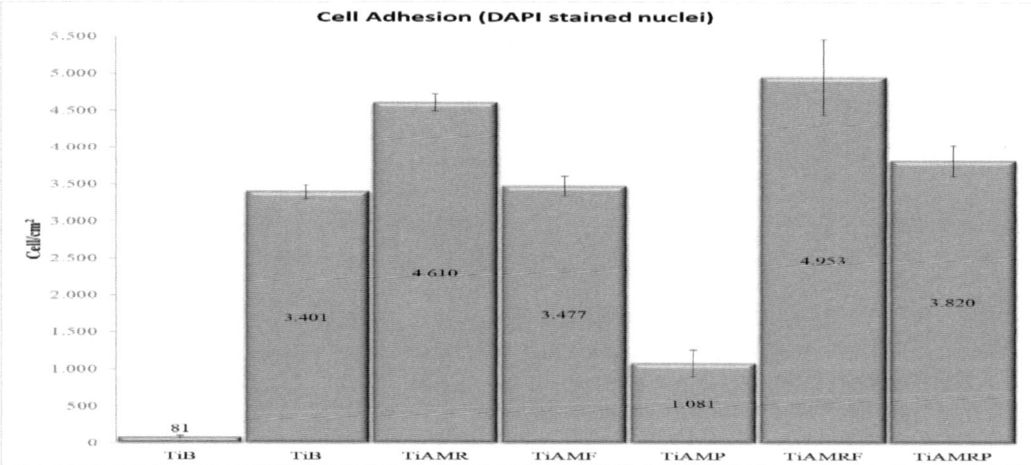

Figure 1. Cell adhesion quantification using staining of the nuclei by DAPI

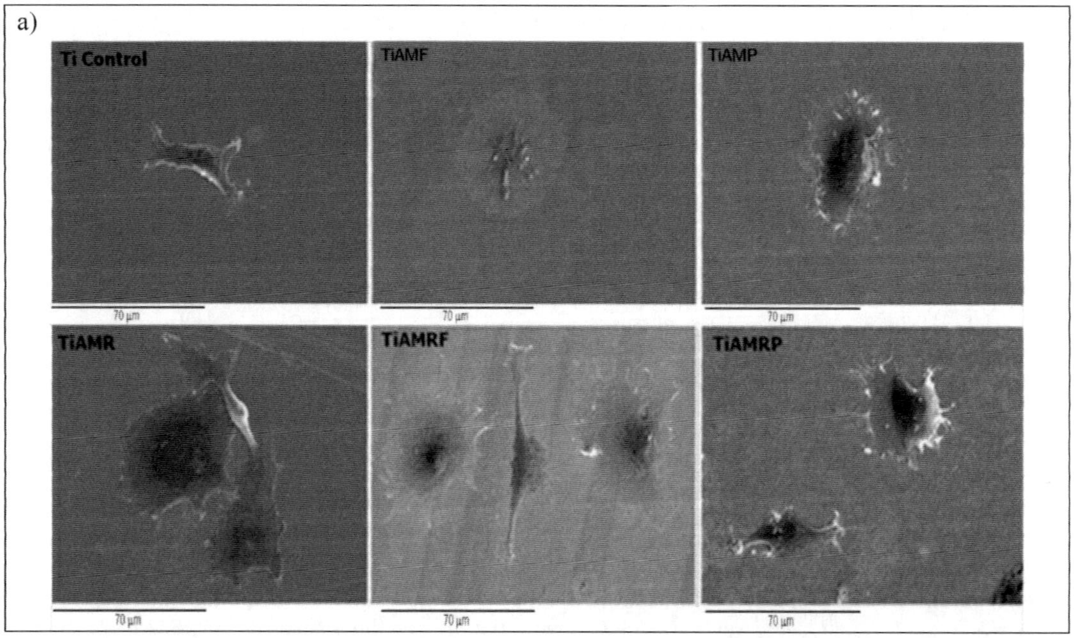

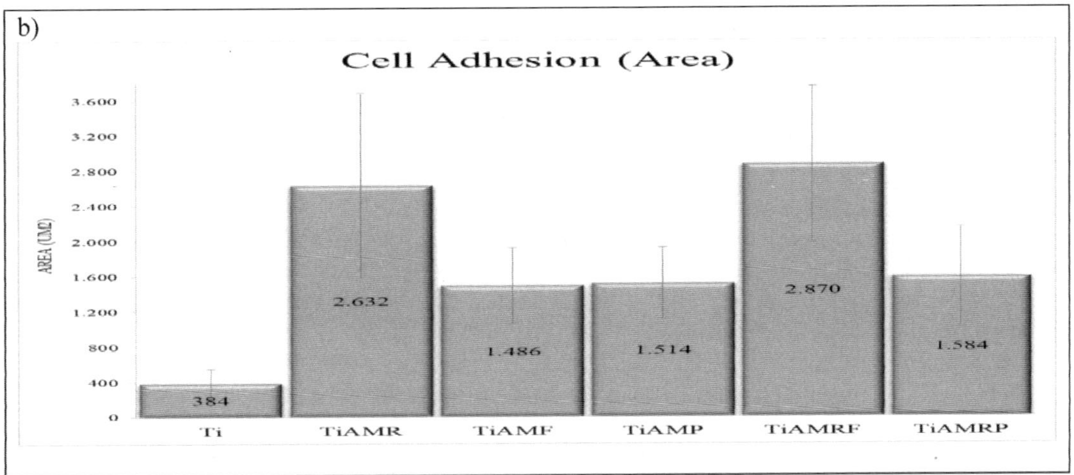

Figure 2. a) Cells Adhered on TiAM by SEM **b)** Area of the Adhered Cells

4. Conclusion

According to the results and in terms of the density of cells adhered, their morphology and the size, the samples with RGD (alone or in mixtures) showed the best results. When correlating the morphology and the area of the cell, we can affirm: 1) although in the untreated sample had a large number of cells, these were contracted, 2) the samples that show the best results are TIAMRF, because they have the greatest extension and marked presence of filopodia. Therefore, we can concluded that our nobel Ti alloy is biocompatible in terms of cellular adhesion, and also that the physical-chemical properties of the surfaces define the size, shape and distribution of the cells.

References

[1] E. Gibon, A.C. Luis, L. Lu, T. Lin, M. Hamadouche, S.B. Goodman, The biological response to orthopedic implants for joint replacement . II: Polyethylene , ceramics , PMMA , and the foreign body reaction, J Biomed Mater Res Part B. (2017) 1685–1691. doi:10.1002/jbm.b.33676.

[2] K. Prasad, O. Bazaka, M. Chua, M. Rochford, L. Fedrick, J. Spoor, R. Symes, M. Tieppo, C. Collins, A. Cao, D.M. Id, K.K. Ostrikov, K. Bazaka, Metallic Biomaterials : Current Challenges and Opportunities, Materials (Basel). 10 (2017) 1–33. doi:10.3390/ma10080884.

[3] J.C. Vanegas Acosta, D. Garzón-alvarado, M. Casale, Interacción entre osteoblastos y superficies de titanio: aplicación en implantes dentales., Rev. Cuba. Investig. Biomédicas. 29 (2010) 51–68.

[4] M.I. Castellanos, J. Guillem-Marti, C. Mas-Moruno, M. Díaz-Ricart, G. Escolar, M.P. Ginebra, F.J. Gil, M. Pegueroles, J.M. Manero, Cell adhesive peptides functionalized on CoCr alloy stimulate endothelialization and prevent thrombogenesis and restenosis, J. Biomed. Mater. Res. - Part A. 105 (2017) 973–983. doi:10.1002/jbm.a.35988.

[5] V. Paredes, E. Salvagni, J.M. Manero, Assessment of Biomimetic Modification on a Novel Low Elastic Modulus Ti–Nb–Hf Alloy for Dental Implants, J. Mater. Sci. Eng. 04 (2015) 193. doi:10.4172/2169-0022.1000204.

[6] L. Salvati, X. Yang, Ultrapasivación de una Aleación con contenido de Cromo, 2011.

[7] V. Paredes, E. Salvagni, E. Rodríguez-Castellón, J.M.J.M. Manero, Comparative Study of Surface Chemical Composition and Oxide Layer Modification upon Oxygen Plasma Cleaning and Piranha Etching on a Novel Low Elastic Modulus Ti25Nb21Hf Alloy, Metall. Mater. Trans. A. 48 (2017) 1–7. doi:10.1007/s11661-017-4144-4.

[8] V. Paredes, E. Salvagni, E. Rodríguez-Castellon, F.J.J. Gil, J.M.M. Manero, Study on the use of 3-aminopropyltriethoxysilane and 3-chloropropyltriethoxysilane to surface biochemical modification of a novel low elastic modulus Ti-Nb-Hf alloy, J. Biomed. Mater. Res. Part B

Appl. Biomater. 103 (2015) 495–502. doi:10.1002/jbm.b.33226.
[9] J.N.-M. Calvo, M. Elices, G. V. Guinea, J. Pérez-Rigueiro, M. Arroyo-Hernández, Stability and activity of lactate dehydrogenase on biofunctional layers deposited by activated vapor silanization (AVS) and immersion silanization (IS), Appl. Surf. Sci. 416 (2017) 965–970. doi:10.1016/j.apsusc.2017.04.123.
[10] X. Chen, P. Sevilla, C. Aparicio., Surface biofunctionalization by covalent co-immobilization of oligopeptides., Colloids Surfaces B Biointerfaces. 107 (2013) 189– 197.
[11] S.L. Bellis, Advantages of RGD peptides for directing cell association with biomaterials., Biomaterials. 32 (2011) 4205–10. doi:10.1016/j.biomaterials.2011.02.029.
[12] A. Shekaran, A.J. García, Extracellular matrix-mimetic adhesive biomaterials for bone repair., J. Biomed. Mater. Res. A. 96 (2011) 261–72. doi:10.1002/jbm.a.32979.
[13] M. Schuler, D.W. Hamilton, T.P. Kunzler, C.M. Sprecher, M. de Wild, D.M. Brunette, M. Textor, S.G.P. Tosatti, Comparison of the response of cultured osteoblasts and osteoblasts outgrown from rat calvarial bone chips to nonfouling KRSR and FHRRIKA-peptide modified rough titanium surfaces., J. Biomed. Mater. Res. B. Appl. Biomater. 91 (2009) 517–27. doi:10.1002/jbm.b.31425.
[14] X.X. Chen, P. Sevilla, C. Aparicio, C. Aparicio., Surface biofunctionalization by covalent co-immobilization of oligopeptides, Colloids Surfaces B Biointerfaces. 107 (2013) 189–197. doi:10.1016/j.colsurfb.2013.02.005.
[15] A.G. Secchi, V. Grigoriou, I.M. Shapiro, R.J. Composto, P. Ducheyne, C.S. Adams, RGDS peptides immobilized on titanium alloy stimulate bone cell attachment , differentiation and confer resistance to apoptosis, J. Biomed. Mater. Res. Part A. (2006). doi:10.1002/jbm.a.
[16] M. Pegueroles Neyra, Interactions Between Titanium Surfaces and Biological Components, Universidad Politécnica de Cataluña, 2009.
[17] H. Sakamoto, Y. Hirohashi, H. Saito, H. Doi, Y. Tsutsumi, Y. Suzuki, K. Noda, T. Hanawa, Effect of active hydroxyl groups on the interfacial bond strength of titanium with segmented polyurethane through gamma-mercapto propyl trimethoxysilane., Dent. Mater. J. 27 (2008) 81–92.
[18] C. Viornery, Y. Chevolot, D. Léonard, B.-O. Aronsson, P. Péchy, H.J. Mathieu, P. Descouts, M. Grätzel, Surface Modification of Titanium with Phosphonic Acid To Improve Bone Bonding: Characterization by XPS and ToF-SIMS, Langmuir. 18 (2002) 2582–2589. doi:10.1021/la010908i.
[19] D.H. Davis, C.S. Giannoulis, R.W. Johnson, T.A. Desai, D. Hal, Immobilization of RGD to < 1 1 1 > Silicon Surfaces for Enhanced Cell Adhesion and Proliferation., Biomaterials. 23 (2002) 4019–27. http://www.ncbi.nlm.nih.gov/pubmed/12162335.
[20] Y. Feng, M. Mrksich, The Synergy Peptide PHSRN and the Adhesion Peptide RGD Mediate Cell Adhesion through a Common Mechanism., Biochemistry. 43 (2004) 15811–21. doi:10.1021/bi049174+.
[21] S.E. Ochsenhirt, E. Kokkolia, J.B.J.B. McCarthy, M. Tirrell, E. Kokkoli, Effect of RGD Secondary Structure and the Synergy Site PHSRN on Cell Adhesion, Spreading and Specific Integrin Engagement., Biomaterials. 27 (2006) 3863–74. doi:10.1016/j.biomaterials.2005.12.012.
[22] P. Schaffner, M.M. Dard, Review: Structure and function of RGD peptides involved in bone biology, Cell. Mol. Life Sci. 60 (2003) 119–132.
[23] C. Satriano, G.M.L. Messina, C. Marino, I. Aiello, E. Conte, D. La Mendola, D. a Distefano, F. D'Alessandro, G. Pappalardo, G. Impellizzeri, Surface immobilization of fibronectin-derived PHSRN peptide on functionalized polymer films-effects on fibroblast spreading., J. Colloid Interface Sci. 341 (2010) 232–9. doi:10.1016/j.jcis.2009.09.046.
[24] D.S.W. Benoit, K.S. Anseth, The Effect on Osteoblast Function of Colocalized RGD and PHSRN Epitopes on PEG Surfaces., Biomaterials. 26 (2005) 5209–20. doi:10.1016/j.biomaterials.2005.01.045.

IOP Conf. Series: Materials Science and Engineering **436** (2018) 012005 doi:10.1088/1757-899X/436/1/012005

Efficient Sparse Representation Response Surface Modelling and Simulation Method for Complex Products

Pu Li[1], Changzheng Huang[1], Songxi Hu[1], Senquan Yang[1], Haoxiang Wen[1] and Lin Zhou[2,*]

[1]School of Physics and Mechanical & Electrical Engineering, Shaoguan University, Shaoguan, China
[2]School of Education, Shaoguan University, Shaoguan, China

*Corresponding author e-mail: zling0126@126.com

Abstract. With the composition and control of modern electromechanical products more and more complex, the time of solving simulation model is longer and longer. The response surface simulation optimization method is an effective method to reduce the simulation time, however, the number of design sampling points of response surface approximation is still large. In this paper, we propose a sparse representation response surface model to accurately reconstruct the source model with small amount sampling points. By means of the sparse representation of the source model on a specific basis, most of the coefficients are zero which can be solved by the equations constructed from small number of sampling. Sparse representation response surface models include sparse response surface and quasi-sparse response surface which are respectively applied to the case where the number of sampling points is greater than and less than the degree of sparse. Sparse response surface runs quicker and quasi-sparse response surface has higher accuracy. Two test functions and one engineering practice problem are employed to compare the performance between sparse representation response surface model and other common response surface models. The results show that the sparse representation response surface model has better performance in approximate accuracy and simulation efficiency.

1. Introduction

With the complex function of modern electromechanical products (such as vehicles, engines, CNC machines and so on), the degree of intelligence is improving, the system composition and control are becoming more and more complicated. Therefore, modelling and simulation technology has been widely used in the design process to improve the overall performance of the product [1].

In the process of products modelling and simulation, simulation models often reflect the multi-disciplinary, non-line and other significant features, which cause a long time to solve the simulation, maybe take as many as 30 to 160 hours [2]. As the product simulation optimization needs to call the simulation process several times, which takes longer time, it is difficult to achieve the objective of product optimal design. Therefore, the efficient simulation of complex products has become an urgent problem in the process of engineering product design.

Because the complex product model has a long computation time and the simulation model may not be a complete explicit mathematical equation, the existing method usually regards the simulation model as a black box function, and uses the optimization algorithm to search globally in the design space to expect to reduce the call times of black box function [3]. Among those methods, the response surfaces simulation optimized method based on computer experiment design is an effective method [4]. For the source model with unknown shape in advance, the response surface of the approximation target is constructed by random sampling and target estimation in the design space, and the accuracy of the response surface is improved by increasing adaptive sampling points. The following response surface models are mainly used at present[2]: (1) Polynomial Response Surface (PRS), (2)Multivariate Adaptive Regression Splines (MARS), (3) Kriging (Ordinary Kriging (OK) and Blind Kriging (BK)), (4) Radial Basis Functions (RBF) and Extended Radial Basis Functions (ERBF), (5)Support Vector Regression (SVR). By means of iterative adaptive sampling and response surface approximation in the design space, these response surface simulation optimization methods establish the intrinsic function relation between design variable and the objective function value to search the optimal design point quickly and improve the simulation optimization efficiency. However, when the response surface shape is more complex, the number of sampling points used to construct the response surface and the number of black box function's calling time will be correspondingly more.

Compressed sampling theory is a new theoretical framework of information acquisition and processing proposed in the field of signal processing [5; 6]. Based on the sparse representation, the signal (or image) can be acquired and reconstructed with high fidelity through a very small amount of non-adaptive sampling. In the simulation optimization process, the target (or constraint) estimate in the design space is similar with the image and signal, and the compression sampling can be used as a new response surface constructing method which we call sparse representation response surface (SRRS). We first construct a set of function bases, if the source model can be sparsely represented on the set of function basis (coefficients are mostly zeros), then the non-zero coefficients can be solved only by a small number of sampling points, so as to build sparse or quasi-sparse response surface which can reduce the calculation cost and improve the simulation optimization efficiency.

The rest of the paper is organized as follow: in section 2, compressed sampling method is introduced; we introduce the method of SRRS include sparse response surface and quasi-sparse response surface corresponding applied to the number of sampling points is greater than and less than the degree of sparse degree in section 3 and section 4; the effect of SRRS is shown and verified in section 5; at last, the conclusion is given.

2. Compressed sampling method

Consider a *n*-dimension signal $\boldsymbol{x} \in \boldsymbol{R}^{n\times 1}$ expended in a set of orthogonal basis $\{\varphi_i\}_{i=1}^{n}$, which is:

$$\boldsymbol{x} = \sum_{i=1}^{n} \theta_i \varphi_i. \tag{1}$$

Written in matrix form:

$$\boldsymbol{x} = \boldsymbol{\Phi}\boldsymbol{\theta}, \tag{2}$$

where $\boldsymbol{\Phi} = [\varphi_1, \varphi_2, \cdots, \varphi_n] \in \boldsymbol{R}^{n\times n}$ is the sparse matrix (also known as dictionary), φ_i is the atom, $\boldsymbol{\theta} = [\theta_1, \theta_2, \cdots \theta_n]^T \in \boldsymbol{R}^{n\times 1}$ is expansion coefficients vector. Assuming that the majority of the elements in the coefficient vector $\boldsymbol{\theta}$ are zero, or close to zero, that is, the number of nonzero elements $k \ll n$, then it can be said that the signal $\boldsymbol{x}$ can be sparse on the basis $\boldsymbol{\Phi}$. k is called sparse degree which means the number of nonzero elements. Compressed observing the signal with an observed matrix $\boldsymbol{\Psi} \in \boldsymbol{R}^{m\times n}$ which is not related to the dictionary $\boldsymbol{\Phi}$:

$$\boldsymbol{y} = \boldsymbol{\Psi}\boldsymbol{x}. \tag{3}$$

We can get m linear observation responses $\boldsymbol{y} \in \boldsymbol{R}^{m}$. Recovering the signal $\boldsymbol{x}$ from compressed observation response $\boldsymbol{y}$ is a problem of solving linear equations. However, since $m < n$, Eq.(3) is invalid linear equations which has infinite solutions. By combining Eq.(2) and Eq.(3), we obtain:

$$\boldsymbol{y} = \boldsymbol{\Psi}\boldsymbol{x} = \boldsymbol{\Psi}\boldsymbol{\Phi}\boldsymbol{\Theta} = \boldsymbol{A}\boldsymbol{\Theta}. \tag{4}$$

Although Eq.(4) is still invalid linear equations, since coefficient vector $\boldsymbol{\Theta}$ is k sparse, we can recover the coefficients $\boldsymbol{\Theta}$ from compressed observation response $\boldsymbol{y}$ by a mathematical method, then obtain signal $\boldsymbol{x}$ from $\boldsymbol{x} = \boldsymbol{\Phi}\boldsymbol{\theta}$. Literature [5] [6] pointed out that the correlation between observation matrix and sparse basis should be low, that is, to meet the condition of Restricted Isometry Property

(RIP). So that we can solve the sparse representation coefficients and reconstruct the source model if the number of sampling points met a certain condition.

Solving the sparse representation coefficients based on the observational system equation model Eq.(4) is essentially an optimization problem that search the easiest solution with the constraint of the equation model, that is to solve Eq.(5):

$$\min\|\boldsymbol{\Theta}\|_0 \quad s.t.\, \boldsymbol{y} = \boldsymbol{\Psi x} = \boldsymbol{A\Theta}, \tag{5}$$

where $\|\boldsymbol{\Theta}\|_0$ is the number of nonzero elements in $\boldsymbol{\Theta}$. It can be found that solving Eq.(5) is a combinatorial optimization problem of NP hard.

At present, the method of greedy and relaxation can be used to solve this problem. The greedy algorithm is represented by matching tracing [7] (MP), orthogonal matching tracing [8] (OMP), generalized orthogonal matching tracing [9] (GOMP), etc. The greedy algorithm is easy to use, and its algorithm complexity is only O(*kmn*), however, it is less adaptive to noise. The relaxation algorithm, such as FOCUSS (Focal Underdetermined System Solver) [10] and BP (Basis Pursuit) algorithm, is computationally larger than the greedy algorithm, but with higher accuracy and more adaptive to noise. The iterative threshold algorithm [11] greatly improves the accuracy by iteration, but the number of iterations is higher and the convergence rate is late.

From the compressed sampling method, it can be found that the higher the representation sparsity of response surface (the smaller the sparse degree *k*), the less the nonzero coefficient of response surface on the representation basis, the smaller the number of sampling points needed for the response surface approximation. The number of calling to the black box function will be less, and the optimization efficiency of the product simulation will be higher. Therefore, we propose a constructing method of sparse representation response surface model in this paper, which includes sparse response surface (SRS) model constructed based on sparse representation theory and quasi-sparse response surface (QSRS) model to apply in the case of source model is not strictly sparse on the basis.

3. Sparse response surface

3.1. The mathematical model

SRS is based on the polynomial response surface model, which is represented as linear model:

$$\hat{\boldsymbol{y}}(\boldsymbol{x}) = \sum_{i=1}^{p} \theta_i \varphi_i(\boldsymbol{x}). \tag{6}$$

It also can be written as matrix form

$$\hat{\boldsymbol{y}}(\boldsymbol{x}) = \boldsymbol{\Phi\Theta}, \tag{7}$$

where $\boldsymbol{x} = [x_1 \cdots x_m]$ is the design sampling point, and m is the number of variables. The atoms of SRS are constructed with $\{\varphi_i(\boldsymbol{x})\}_{i=1,2,\cdots p}$,and p is the number of atoms. $\boldsymbol{\Phi}$ is called dictionary, which is a set of basic functions. $\{\theta_i\}_{i=1,2,\cdots p}$ are the atoms coefficients and $\boldsymbol{\Theta}$ is the coefficient vector. We select polynomials functions as the basic atoms of SRS, and atoms is tensor product of the whole univariate polynomials. The univariate polynomials are decided by the definition of polynomials and the index vector. Thus , $\varphi_i(\boldsymbol{x})$ may be defined as

$$\varphi_i(\boldsymbol{x}) = L(\boldsymbol{x}, \boldsymbol{\eta}^{(i)}) = \prod_{j=1}^{m} l_j\left(x_j, \eta_j^{(i)}\right), i = 1, \cdots, p. \tag{8}$$

where $\boldsymbol{\eta}^{(i)} = \left[\eta_1^{(i)}, \cdots, \eta_m^{(i)}\right]^T$ is the exponent vector of $\varphi_i(\boldsymbol{x})$, presented the order of polynomial. $L(\boldsymbol{x}, \boldsymbol{\eta}^{(i)})$ is the polynomial of basis, which is calculated through taking variable $\boldsymbol{x}$ into the polynomial definition and the order number. $l_j\left(x_j, \eta_j^{(i)}\right)$ is the $\eta_j^{(i)}$-order variable polynomial with respect to x_j.

Given a set of sampling points $\boldsymbol{x} = [x^{(1)}, \cdots, x^{(n)}]^T$, $x^{(k)} \in R^m, k = 1,2,\cdots,n$, and the corresponding actual response $\boldsymbol{y} = [y^{(1)}, \cdots, y^{(n)}]^T$, then the dictionary can be defined as

$$\boldsymbol{\Phi} = \begin{bmatrix} \varphi_1(x^{(1)}) & \cdots & \varphi_p(x^{(1)}) \\ \vdots & \ddots & \vdots \\ \varphi_1(x^{(n)}) & \cdots & \varphi_p(x^{(n)}) \end{bmatrix}. \tag{9}$$

Thus, the model of sparse response surface may be noted as

$$\hat{\boldsymbol{y}}(\boldsymbol{x}) = \boldsymbol{\Phi}\boldsymbol{\Theta}, \quad s.t. \ \|\boldsymbol{\Theta}\|_0 \leq s. \tag{10}$$

3.2. Dictionary construction

Dictionary determined the expression ability of response surface in SRS. In the theory of compression sensing, dictionary may be constituted of a set of orthogonal sparse representation basis like Fourier, discrete cosine, wavelets etc. A set of Legendre polynomial function is employed as basis in the SRS. And the Legendre polynomial is obtained by the circular definition. Assuming that $L_0(x) = 1, L_1(x) = x$,, define the following function as

$$(n+1)L_{n+1}(x) = (2n+1)L_n(x) - nL_{n-1}(x), \ n = 1,2,\cdots. \tag{11}$$

where the subscript n represents the order of Legendre polynomials. The reason why selected Legendre polynomials as basis function is that it can be orthogonal with respect to $\ell 2$ norm meaning in the interval [-1, 1], and this excellent feature may be extended from univariate to multivariate. In addition, Legendre polynomial is convenient to be constructed and calculated.

In addition, the choice of dictionary scale, which determines the numbers of atoms, need to be trade off between the dictionary expression ability and the solution stability. If dictionary scale is too small, it will be not enough to embody the model characteristics in sampling points and the function relationship. Whereas the dictionary is too big, calculating coefficient will be a serious underdetermined problem, and solution stability will become poorer. According to the literature [12], the number of sampling points n should be at least 15% of the number of atoms p. In this paper, we set $p = 6n$. In the meanwhile, to avoid overfitting and reduce computation, it prefers to select low-order function when choosing the Legendre polynomial function to construct atoms To guarantee the total order is as small as possible, the atomic structure of multivariate model also obey this criterion.

The choice of sampling position commands that the sampling may express characterize of source model to the greatest extent. In general, sampling points should be uniform distribution on the entire design feasible region. Uniform design (UD) and Latin hypercube design (LHD) are meet the demand, since both of them can divide design space into n sampling interval uniformly, and choose one sampling points in each interval. The difference is that sampling position of UD is fixed in each interval, while sampling position of LHD is random in the corresponding interval. In the SRS, LHD is employed for sampling, because the randomness in LHD can shrink off the correlation of atoms in the dictionary and facilitate the resolution of atoms expressing different characteristics.

Due to the definition intervals of Legendre polynomials function is not the same as the design space of design model, it is necessary to map sampling points of design space to the definition intervals of Legendre polynomials function $[-1,1]^m$. Then, calculate value of atoms by taking sampling points $\boldsymbol{x}$ into Legendre polynomials function, and dictionary $\boldsymbol{\Phi}$ can be constructed.

3.3. Construction of sparse response surface

The sparseness of source model in polynomial basis function is unknown. To build the model of SRS, the coefficients of the basis function can be calculated through the following function.

$$\boldsymbol{\Theta} = min\|\boldsymbol{\Theta}\|_0 \, . \ s.t. \|\boldsymbol{y} - \boldsymbol{\Phi}\boldsymbol{\Theta}\|_2^2 \leq \epsilon. \tag{12}$$

Due to the object $\ell 0$ norm is a nonconvex problem, we can relax the sparse object condition to $\ell 1$ norm. Eq.(12) is translated to :

$$\boldsymbol{\Theta} = min\|\boldsymbol{\Theta}\|_1 \, . \ s.t. \|\boldsymbol{y} - \boldsymbol{\Phi}\boldsymbol{\Theta}\|_2^2 \leq \epsilon. \tag{13}$$

$\ell 1$ norm is a convex problem, which is the closest to $\ell 0$ norm, and it's Lagrange multiplier can be

$$\widehat{\boldsymbol{\Theta}}(\lambda) = \arg\min_{\boldsymbol{\Theta}} \ \|\boldsymbol{y} - \boldsymbol{\Phi}\boldsymbol{\Theta}\|_2^2 + \lambda\|\boldsymbol{\Theta}\|_1. \tag{14}$$

Eq.(14) is a famous regression model names least absolute shrinkage and selection operator (LASSO), which may provide both variable selection and coefficient shrinkage functions, is commonly applied in statistics. Variable selection can be used to generate the model of sparse response surface, and coefficient shrinkage can achieve better prediction performance through compromising the model training accuracy. The LASSO model can be solved by least angle regression (LAR). And author of literature [13] provides a MATLAB toolbox for using to calculate.

The process of constructing SRS is shown as following Table 1.

Table 1. The process of construction sparse response surface.

Step1:	**Generate sampling points.** LHD method is applied to sampling with the lhsdesign of Latin hypercube function in MATLAB. Since the sampling interval of lhsdesign function is [0,1]m, it need to be mapped the sampling points towards design space of source model.
Step2:	**Estimate objective value.** Obtain the objective value $\boldsymbol{y}$ of corresponding sampling points by simulating the black box function of source model.
Step3:	**Construct dictionary.** The sampling points sequence is mapped to Legendre polynomial definition interval[-1,1]m. Then obtain index vector $\{\boldsymbol{\eta}^{(i)}\}_{i=1,2,\cdots p}$, and calculate the value of dictionary.
Step4:	**Calculate coefficients.** Solve Eq. (13) through invoking the LASSO solver solution.
Step5:	**The end of construction.**

4. Quasi-sparsity response surface

The ability of LASSO to provide sparse models has been demonstrated in the field of statistics, but the number of atoms chosen by LASSO cannot exceed the number of rows or columns of the dictionary. For the sparse representation of the response surface, if the number of sampling points n is greater than the sparse degree k, then LASSO will select k atoms to construct a sparse response surface, which can be seen as the precise surrogate model of the source model. However, the source model is complicated in real application, its sparse degree based on the selected basis functions may be not small enough, and the increase of the sampling points will consume the expensive calculation cost, then the number of sampling points is less than sparse degree. Since the model performance is limited in the number of atoms selected by LASSO, we consider the Elastic net regression to construct a quasi-sparsity response surface model which will be more stable by selecting more atoms.

4.1. Definition of QSRS model

Consider the cost function

$$\mathcal{L}(\lambda_1,\lambda_2,\theta) = \|\boldsymbol{y}-\boldsymbol{\Phi\Theta}\|^2+\lambda_1\|\boldsymbol{\Theta}\|_1+\lambda_2\|\boldsymbol{\Theta}\|_2^2, \tag{15}$$

where

$$\|\boldsymbol{\Theta}\|_1=\sum_{j=1}^{p}|\theta_j|,\ \|\boldsymbol{\Theta}\|_2^2=\sum_{j=1}^{p}\theta_j^2.$$

The Elastic net estimator $\widehat{\Theta}$ is the minimize of Eq.(14):

$$\widehat{\boldsymbol{\Theta}} = argmin_{\boldsymbol{\Theta}}\{\mathcal{L}(\lambda_1,\lambda_2,\theta)\}. \tag{16}$$

From the Eq.(15), we can see that the elastic net is a regularized regression method that linearly combines the $\ell 1$ and $\ell 2$ penalties of LASSO regression and ridge regression.

4.2. Solution of Elastic net regression

Theorem. Given data set (**y**,**X**) and (λ_1, λ_2), the elastic net estimation is

$$\boldsymbol{\Theta}(elastic\ net) = argmin_{\boldsymbol{\Theta}}\,\boldsymbol{\Theta}^T\left(\frac{\boldsymbol{X}^T\boldsymbol{X}+\lambda_2\boldsymbol{I}}{1+\lambda_2}\right)\boldsymbol{\Theta}-2\boldsymbol{y}^T\boldsymbol{X\Theta}+\lambda_1|\boldsymbol{\Theta}|_1. \tag{17}$$

It is easy to see that

$$\boldsymbol{\Theta}(lasso) = argmin_{\boldsymbol{\Theta}}\,\boldsymbol{\Theta}^T(\boldsymbol{X}^T\boldsymbol{X})\boldsymbol{\Theta}-2\boldsymbol{y}^T\boldsymbol{X\Theta}+\lambda_1|\boldsymbol{\Theta}|_1. \tag{18}$$

The proof of theorem is simple, and we omit it here. We can find there is a convenience for elastic net that the estimation process can be turned to the LASSO solution from above subsection. We introduce a lemma as below.

Lemma. Given data set (**y**, **X**) and (λ_1, λ_2), define an artificial data set (**y***, **X***) by

$$\boldsymbol{X}^*_{(n+p)\times p}=(1+\lambda_2)^{-\frac{1}{2}}\begin{pmatrix}\boldsymbol{X}\\ \sqrt{\lambda_2}\boldsymbol{I}\end{pmatrix},\ \boldsymbol{y}^*_{(n+p)}=\begin{pmatrix}\boldsymbol{y}\\ \boldsymbol{0}\end{pmatrix}. \tag{19}$$

Let $\gamma=\lambda_1/\sqrt{(1+\lambda_2)}$, $\Theta^*=\sqrt{(1+\lambda_2)}\Theta$. Then LASSO question is equaled with

$$\mathcal{L}(\gamma,\boldsymbol{\Theta}) = \mathcal{L}(\gamma,\boldsymbol{\Theta}^*) = |\boldsymbol{y}^* - \boldsymbol{X}^*\boldsymbol{\Theta}^*|^2 + \gamma|\boldsymbol{\Theta}^*|_1. \quad (20)$$

Let $\Theta^* = \text{argmin}_{\Theta^*}\{\mathcal{L}(\gamma,\Theta)\}$, then

$$\boldsymbol{\Theta}(elasticnet) = \sqrt{1+\lambda_2}\boldsymbol{\Theta}^*. \quad (21)$$

We omit the proof for it is just simple algebra. For the fixed λ_2, the elastic net problem can be changed to a simple LASSO problem which can be solved by the "LAR-EN" solver. In Ref [14], the author gives a "LAR-EN" solver that can directly find Elastic net solutions for fixed λ_2.

4.3. Construction of QSRS

There are two main differences between the construction of QSRS and SRS construction: the choice of parameter λ_2; the sampling mode is changed from LHD sampling to UD sampling.

In the construction of the QSRS, the parameter λ_2 is determined by the method of cross-validation. The specific flow is shown in Table 2.

Elastic net will select a group of similar atoms, regardless of the correlations between atomics, and the sampling points of UD is fixed, then, the response surface is fixed, the reconstruction effect is stably. Thus, UD sampling is employed in the constructing of quasi-sparsity response surface.

Table 2. The process of sparse response surface construction.

Step1:	Generate sampling points by Uniform sampling.
Step 2:	Get the responding value $\boldsymbol{y}$ of the sampling points.
Step 3:	Divide the sampling points into K sets. (K is priority to use 10, or other values that can divide the number of sampling points)
Step 4:	Respectively, fix$\lambda_2$to0, 0.0001,0.001,0.01,0.1,1and 10, run steps 5 to 7.
Step 5:	K-1 copies of the sampling points and responding values are taken as the training set, and the remaining one is the prediction set. The Quasi-sparse response surface model is constructed by using the LAR-EN solver with the training set, the prediction accuracy of the model is evaluated on the prediction set.
Step 6:	Repeat step 5 to ensure that each set is used for predicting.
Step 7:	Calculate the mean of the predictor error as the prediction ability of every λ_2.
Step 8:	Take $\lambda_2$with thebest prediction ability.
Step 9:	Construct the QSRS using the LAR-EN solver with the training set and the determined λ_2 .

5. Numerical experiments

In order to verify the appreciate effect of SRS and QSRS, 2 simulation test functions and one practical engineering design problem are chosen in this section as the object to be reconstructed, and some commonly used response surface models are chose as a comparison.

5.1. Experiments problems

1) Test function 1 (10 variables)

$$f(x) = \sum_{i=1}^{m}[0.6 + sin\left(\frac{16*x_i}{15} - 1\right) + sin\left(\frac{16*x_i}{15} - 1\right)^2 + sin(4*\left(\frac{16*x_i}{15} - 1\right))], \quad (22)$$

where $x_i \in [-1,1]$, *i*=1,2,3,…,*m*, *m*=10.

2) Test function 2 (20 variables)

$$f(x) = 1 + \sum_{i=1}^{m}\{8\sin^2[7(x_i - 0.9)]^2 + 6\sin^2[14(x_i - 0.9)]^2 + (x_i - 0.9)^2\}, \quad (23)$$

where $x_i \in [-500,500]$, *i*=1,2,3,…,*m*, *m*=20.

3) Engineering design problem

Reference [15] provides a few design studies based on finite element analysis and natural experiment design of composite sheet structures. The analysis process is complicated in the optimization process. The response surface method can provide an approximate model for this kind of problem, save the computation time and improve the simulation optimization efficiency. We choose a global deflection structure optimization example of a square composite sheet as the validated case of SRRS model. The specific description and data of the case can be downloaded from http://www.cs.rtu.lv/jekabsons/datasets.html. The optimization target of this case is decided by six variables, and the variable description and range of values are shown in Table 3.

Table 3. Variables description of engineering design problem.

Variable	Description	Minimum	Maximum	Units
L	Panel length	3	7	m
h	Panel height	4	16	mm
t1	Top and bottom plate thickness	2	4	mm
t1	Core stiffener thickness	1.5	4	mm
kh	Core stiffener spacing factor	1.5	4	
n	Symmetrical number of core stiffeners	2	6	

5.2. Accuracy assessment criteria

The following metrics are used to assess the generalization capabilities of the response surface models compared in this work:

1) Root mean squared error(RMSE)

The RMSE is given by

$$RMSE = \sqrt{\frac{1}{n}\sum_{k=1}^{n}(y^{(k)} - \hat{y}^{(k)})^2}. \tag{24}$$

where $y^{(k)}$ is the actual response and $\hat{y}^{(k)}$ is the predicted response at the k-th test point, n is the number of test points.

2) Maximum absolute error(MAE)

The maximum absolute error is given by

$$MAE = max\left|y^{(k)} - \hat{y}^{(k)}\right|, k = 1,2,\cdots,n. \tag{25}$$

3) Correlation coefficient (R)

The correlation coefficient between actual response and predicted response is expressed as

$$R = \frac{\left(\sum_{k=1}^{n}(y^{(k)}-\bar{y})(\hat{y}^{(k)}-\bar{\hat{y}})\right)}{\left(\sqrt{\sum_{k=1}^{n}(y^{(k)}-\bar{y})^2}\sqrt{\sum_{k=1}^{n}(\hat{y}^{(k)}-\bar{\hat{y}})^2}\right)}. \tag{26}$$

where $\bar{y}$ is the mean of actual response and $\bar{\hat{y}}$ is the mean of prediction responses.

5.3. Models settings

SRS and QSRS are compared with the models listed in introduction section including PRS, OK, BK, RBF, ERBF, MARS, and SVR. The parameters of these models are listed in Table 4.

Table 4. The detail settings of different surrogate models.

Surrogate model	Details
PRS	p=2
OK	A constant regression function and a Gaussian correlation model are employed in the mode. In all cases, $\theta_0=1_{m*1}$, and $0.1\leq \theta_i \leq 20$, for i = 1, … ,m, where *m* and $1_{m*1}$are the number of variables and the vector whose entries are all equal to 1, respectively
BK	The same as OK.
RBF	The form of basis functions is the multi quadric function and we set c = 0.9.
ERBF	The parameter n is set equal to 2 and the parameter c is set to be approximately 1/3 of the average domain size. The remaining options are the same as RBF.
MARS	The piecewise-cubic regression functions are adopted in this approach and the stopping criterion for the forward phase is set to 1e-4. Additionally, the maximal number of basis functions in the forward model building phase is 81 and the Generalized Cross-Validation penalty per knot is 3.
SVR	We choose the RBF kernel function which is set with $\sigma = 4$ and the e-insensitive loss function. The parameters C, e are selected as $C = 100\max(\lvert\bar{y} + 3\sigma_y\rvert, \lvert\bar{y} - 3\sigma_y\rvert)$ and $\varepsilon = \delta_y/\sqrt{ns}$, respectively. Here, $\bar{y}$ and σ_y the mean value and the standard deviation of the function values at the training points

5.4. Experiment settings

For all validated objects, the sampling points and test points are generated by LHD in the design domain, and the sampling points are used to construct the response surfaces. The test points are used to test the model approximation accuracy. To reduce the effect of random sampling on the test results, the test is to be repeated 300 or 100 times. The average of the accuracy criteria is used to measure the approximation of the response surface model. Detailed can be seen in Table 5.

Table 5. Numerical setup for test problems.

Test problem	NO. of variables	No. of training points	No. of test points	No. of test times
Test function 1	10	100	4500	100
Test function 2	20	120	5000	100
Engineering problem	6	50	450	300

5.5. Experiments analysis and discussion

5.5.1. RMSE. Table 6 shows the comparison of the average RMSE results of different response surface models. It can be seen that the RMSEs of both SRS and QSRS models are significantly better than other common response surface models in this test. From the test results of test function 1, all tested response surface models' RMSEs are low. The performance of QSRS is the best, which is 11.6% higher than that of SRS. In addition to SRS and QSRS, ERBF and MARS also have a good approximation effect. For test function 2, because the number of variables reaches 20 whereas the number of sampling points is only 120, so most of the response surface models can not be a good object to the reconstruction of the approximation, however, both SRS and QSRS are doing well. For the engineering design problem, due to the small number of variables, all the models are doing well, SRS and QSRS have the advantage relative to other models. In general, SRRS has a significant advantage in approximating RMSE, and this advantage is more pronounced in the case of multi-variable and less sampling points. In response to SRS and QSRS, the performance of QSRS is better than SRS. Because when the parameter $\lambda_2 = 0$, the QSRS degrades to the SRS, so the QSRS has performance that is not weaker than the SRS. However, it must be noted that the construction of the QSRS is more expensive than the SRS due to the cross validation process

Table 6. Mean of RMSEs for different surrogate models, the optimal value in each column is shown in bold for ease of comparison.

Surrogate model	Test function 1	Test function 2	Engineering design problem
PRS	3.4667	9.658e5	0.0065
OK	2.3096	3.332e5	0.0122
BK	4.0665	1.786e5	0.0061
RBF	2.2124	1.711e5	0.0143
ERBF	1.6026	5.013e3	0.0203
MARS	1.6193	8.776e4	0.0081
SVR	2.8475	4.176e5	0.0060
SRS	1.4967	34.821	0.0041
QSRS	**1.3415**	**33.826**	**0.0037**

5.5.2. MAE. Table7 compared the mean MAE results of all models. It can be seen that the trend of MAE is consistent with the trend of RMSE from the results. SRS and QSRS are also better than the other response surface models, and the advantage is more obvious in the case of multi-variable and less sampling points such as test function 2 which have 20 variables.

IOP Conf. Series: Materials Science and Engineering **436** (2018) 012005 doi:10.1088/1757-899X/436/1/012005

Table 7. Mean of MAEs for different surrogate models, the optimal value in each column is shown in bold for ease of comparison.

Surrogate model	Test function 1	Test function 2	Engineering design problem
PRS	15.9729	4.231e6	0.0401
OK	8.3281	1.303e6	0.0834
BK	18.3734	8.611e5	0.0378
RBF	8.4164	7.504e5	0.0834
ERBF	6.9643	1.990e4	0.0963
MARS	7.4095	3.345e5	0.0537
SVR	11.7081	1.827e6	0.0478
SRS	6.2611	146.447	0.0330
QSRS	**6.0611**	**140.756**	**0.0313**

5.5.3. Correlation coefficient. Table 8 shows the comparison of the mean of correlation coefficient results for different surrogate models. From the results of test function 1, the performance of the QSRS is still the best, the performance of ERBF and MARS are also good. It is worth noting that the performance of SRS is not as good as ERBF and MARS, the reason should be the sparse degree of test function 1 is greater than the number of sampling points, then the number of selected atoms is too small and the response surface is not enough stable. For the test function 2, SRS, QSRS, and MARS all behave well, and the mean R of MARS is even higher than the SRS. For the engineering design problem, the performance of QSRS and SRS response are in the top two. It can be seen that although there are other response surface models may be better than SRRS in the correlation coefficient, but on the whole, both SRS and QSRS play a stable and excellent performance.

Table 8. Mean of Rs for different surrogate models, the optimal value in each column is shown in bold for ease of comparison.

Surrogate model	Test function 1	Test function 2	Engineering design problem
PRS	0.2480	0.3309	0.8838
OK	0.0179	1.81e-8	0.4748
BK	0.1864	0.8543	0.8832
RBF	0.4125	0.9365	0.4667
ERBF	0.8262	0.9399	0.4260
MARS	0.8125	0.9633	0.7986
SVR	0.2454	0.0778	0.8982
SRS	0.7758	0.9477	0.9491
QSRS	**0.8369**	**0.9499**	**0.9621**

6. Conclusions

In this paper we proposed a sparse representation response surface include SRS and QSRS based on compressed sampling method. The Legendre polynomial function is employed as the sparse basis, the SRS and QSRS models are constructed by LASSO or Elastic net regression method and LHD or UD sampling. The SRS and QSRS are compared with other commonly used response surface model in the numerical experiments of two standard test functions and one practical engineering problem. After the accuracy criteria evaluation of RMSE, MAE, and R, the results show that both SRS and QSRS have better approximation ability than other commonly used response surface models.

The conclusions can be given as below: 1)The sparse representation response surface can be used to approximate the simulation object, and only a small amount of sampling can obtain effective simulation and optimization efficiency, 2)Sparse representation response surface has a better approximation effect compared to the other commonly used response surfaces tested in this paper, and this advantage is more obviously in the case of multi-variable and less sampling.

Acknowledgments

This work was financially supported by Natural Science Foundation of Guangdong Province of China under Project 2015A030310510 and 2016KQNCX156, and the Science and Technology Program of Shaoguan City of China (441-99000318 and 441-99000319).

References

[1] Torry-Smith J M, Qamar A, Achiche S, et al. Challenges in Designing Mechatronic Systems[J]. Journal of Mechanical Design, 2013, 135(1): 184-194.

[2] Wang G G, Shan S. Review of Metamodeling Techniques in Support of Engineering Design Optimization[C]. ASME 2006 International Design Engineering Technical Conferences and Computers and Information in Engineering Conference, 2006: 415-426.

[3] Hong L J, Nelson B L. A brief introduction to optimization via simulation[C]. Winter Simulation Conference, 2009: 75-85.

[4] Romero D A, Amon C H, Finger S. Multiresponse Metamodeling in Simulation-Based Design Applications[J]. Journal of Mechanical Design, 2012, 134(134): 091001.

[5] Candes E J, Wakin M B. An Introduction To Compressive Sampling[J]. IEEE Signal Processing Magazine, 2008, 25(2): 21-30.

[6] Donoho D L. Compressed sensing[J]. IEEE Transactions on Information Theory, 2006, 52(4): 1289-1306.

[7] Mallat S G, Zhang Z. Matching pursuits with time-frequency dictionaries[J]. IEEE Transactions on Signal Processing, 1993, 41(12): 3397 - 3415.

[8] Tropp J A, Gilbert A C. Signal Recovery From Random Measurements Via Orthogonal Matching Pursuit[J]. IEEE Transactions on Information Theory, 2007, 53(12): 4655-4666.

[9] Jian W, Kwon S, Shim B. Generalized Orthogonal Matching Pursuit[J]. IEEE Transactions on Signal Processing, 2012, 60(12): 6202-6216.

[10] Gorodnitsky I F, Rao B D. Sparse signal reconstruction from limited data using FOCUSS: a re-weighted minimum norm algorithm[J]. IEEE Transactions on Signal Processing, 2002, 45(3): 600-616.

[11] Blumensath T, Davies M E. Iterative Thresholding for Sparse Approximations[J]. Journal of Fourier Analysis and Applications, 2008, 14(5): 629-654.

[12] Gribonval R, Cevher V, Davies M E. Compressible Distributions for High-Dimensional Statistics[J]. IEEE Transactions on Information Theory, 2012, 58(8): 5016-5034.

[13] Efron B, Hastie T, Johnstone I, et al. Least Angle Regression[J]. Annals of Statistics, 2004, 32(2): 407-451.

[14] Zou H, Hastie T. Regularization and Variable Selection via the Elastic Net[J]. Journal of the Royal Statistical Society, 2005, 67(2): 301-320.

[15] Kalnins K, Eglitis E, Jekabsons G, et al. 3.2 – Metamodels for Optimum Design of Laser Welded Sandwich Structures[J]. Design Fabrication & Economy of Welded Structures, 2008: 119-126.

Simultaneously Toughening and Strengthening of PLA-b-PCL Copolymer via Pressure-induced-flow Processing at Room Temperature

S Zhu[1,2,*], K Zhang[1,2], Y Hu[1,2], M Xiong[1,2], S Zhang[1,3], M Jiao[1,4] and M Yu[1,2,*]

[1]State Key Laboratory for Modification of Chemical Fibers and Polymer Materials, Donghua University Centre for Civil Aviation Composites, Donghua University, NorthRenmin Rd. 2999, Songjiang District, Shanghai 201620, China;
[2]Shanghai Key Laboratory of Lightweight Composite, Donghua University, Shanghai201620, P. R. China;
[3]Liaoning Engineering Technology Research Center of Function Fiber and Its Composites, Dalian Polytechnic University, Dalian 116034, China
[4]School of Materials & Chemical Engineering, Zhongyuan University of Technology, Zhengzhou 450007, China

*Corresponding authors:zhushu@dhu.edu.cn (Shu Zhu), yumuhuo@dhu.edu.cn (Muhuo Yu)

Abstract. Biodegradable but brittle polylactide (PLA) can be toughened by copolymerizing with poly caprolactone (PCL). However, the tensile strength and Young's modulus dramatically decrease compared with pure PLA. On the other hand, melting processing is normally applied for most plastics at relatively high temperatures and may lead to substantial degradation of polymers like PLA-b-PCL copolymer. In this study, we adopted pressure-induced-flow (PIF) processing to manufacture PLA-b-PCL copolymer specimens at room temperature of ~30 °C. Impact toughness, tensile strength and Young's modulus were simultaneously increased by about 190 %, 151 % and 107 % respectively, when compared with the samples processed via injection moulding. We attributed the remarkable improvements to the multi-layered morphologies generated during PIF-processing which are parallel to the flowing direction. Moreover, the molecular weight and all mechanical properties for PLA-b-PCL were well preserved at such low processing temperature after being reprocessed for ten times.

1. Introduction

Polylactide (PLA), produced from renewable resources, has attracted intensive research attention because of its excellent properties and promising perspective to be alternative of current petroleum-based plastics. However, the application is limited by its intrinsic brittleness, elevated price and degradation during recycling [1-5]. Polycaprolactone (PCL) attracts great interest mainly due to its lower cost and higher toughness than other biopolymers [6-9]. Therefore, through last two decades, toughening PLA by blending or copolymerizing with PCL has become the most promising direction to expand applications of both polymers [10-12]. Previous studies have demonstrated that, generally, with the content of PCL increasing, the toughness of PLA-b-PCL copolymer was improved effectively, but the tensile strength and Young's modulus decreased dramatically [13-17]. For this reason, many studies have been dedicated to modify hierarchical structures by tuning types and fraction of component or solidification of each block [18-22]. However, the contradiction between enhanced toughness and impaired strength and modulus always existed.

Pressure-induced-flow (PIF) processing has been demonstrated to be an effective method to enhance together the toughness and strength of semi-crystalline and amorphous polymer, such as toughening of polypropylene (PP) [23] and high impact polystyrene (HIPS) [24]. In addition, the PIF-processing temperature has been adopted slightly beyond glass transition temperature of the material, but much lower than the melting temperature. Therefore, such relatively low temperature processing method shows great advantage in maintaining properties of degradable polymers, such as PLA-b-PCL. Taniguchi et al [25] revealed that PLA-b-PCL showed room temperature processability under hydraulic pressure of 34.5 MPa without polymer degradation. They proposed a greater hard block composition led to a higher modulus, a higher stress-at-break and a lower strain-to-break. However, they did not mention the effect of either pressure or reprocessing times on mechanical properties of PLA-b-PCL.

In this paper, we employed low temperature pressure-induced-flow (PIF) processing to enhance the toughness of PLA-b-PCL and simultaneously avoid sacrificing of tensile strength and Young's modulus. Moreover, such processing technique can be repeated for more than 10 times without significant degradation and deterioration of mechanical properties for PLA-b-PCL copolymer.

2. Experimental procedure

Materials: PLA-b-PCL copolymer was synthesized according to Ref [20]. Weight average molecular weight (M_w) and number average molecular weight (M_n) is 150,000 and 80,000 g/mol, characterized by GPC. Content of PCL was about 30 wt.%. PCL (M_n ~10000 g/mol), was purchased from Guoyao Chemical Agentia Company of China.

Sample preparation: PLA-b-PCL powder was fitted into the channel-die used in successive PIF-processing [26] (as shown in figure 3a). The PIF-processing was conducted with a compressive equipment at room temperature of about 30°C under a certain pressure (in a range of 150MPa to 550MPa) for 10 minutes. In reproduced samples, PLA-b-PCL specimens were cut into fragments and then formed again by PIF-processing. In all studied samples, no stress whitening or shear band observed through optical microscope observation. Cavitation that frequently observed during stretching was also suppressed due to side constraints according to Ref [27].

Measurements: Notched Izod impact strength test was carried out on a Ceast Pendulum Impact Tester (CEAST RESIL6957 Impactor, Italy), following the standard of ISO-180-2000. Tensile strength test was performed on the Universal Tensile Machine (WDW-20, shanghai, China) at a cross head speed of 20 mm/min (according to ISO 527). Scanning electron microscope (SEM) (JSM-5600LV, JEOL) was used to observe fracture surface of samples. The weight-average molecular weight and polydispersity were determined by Gel permeation chromatography (GPC), using THF as a mobile phase at a flow rate of 1.0 ml/min, and the column temperature was 35 °C.

3.Results and discussion

3.1 The effect of pressure on mechanical properties and morphologies

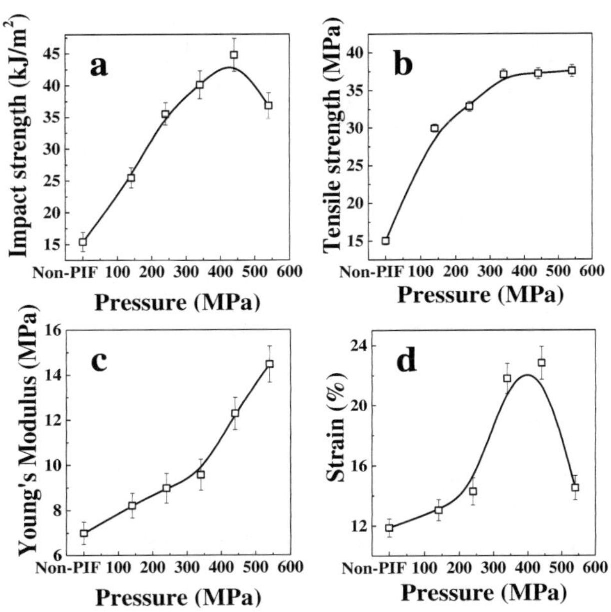

Figure 1. Impact of PIF-processing pressure on mechanical properties of PLA-b-PCL (30 wt. % PCL content) copolymer with and without PIF-processing: (a) impact strength, (b) tensile strength, (c) Young's modulus and (d) strain. (The samples were all processed at 30 °C for 10 min.)

The mechanical properties of PLA-b-PCL copolymer with and without PIF-processing are shown in figure 1. In general, there is a conflict between toughness and strength, which means that by a single processing, the toughness increasing with loss of tensile strength [27]. Here, the PIF-processing method is suggested to be a promising technique that can enhance toughness and strength simultaneously, which is extremely essential for expanding application field of biomaterials. It can be seen that the toughness of PLA-b-PCL copolymer increased by 190 % indicated by the impact test (as shown in figure 1a). Simultaneously, tensile strength, Young's modulus and tensile strain increased by 151 %, 107 % and 92 % respectively (as shown in figure 1b, c, d) compared with PLA-b-PCL specimens processed via injection moulding. In addition, PIF-processing was performed at lower processing temperature (~30 °C), which may be important to avoid significant degradation problem during processing compared with traditional melt processing.

Impact post-fractured surface of PLA-b-PCL copolymer (30 wt. % PCL content) with and without PIF-processing were studied by SEM, as shown in figure 2. The fractured surface of injection molded samples exhibited smooth and isotropic morphologies. However, after PIF-processing, the fractured surface showed rougher features compared with the original as shown in figure 2a. Many of oriented grooves along flow direction (FD) can be found, furthermore grooves became deeper and longer with higher pressure as shown in figure 2(b, c, d), which indicated that more energy was dissipated upon impact. Such oriented grooves along FD implied that oriented structures were generated during PIF-processing. According to our previous researches on PP [26] and PLA/PEG blends [28], with PIF-processing, crystalline lamellae fractured and subsequently reoriented along FD direction.

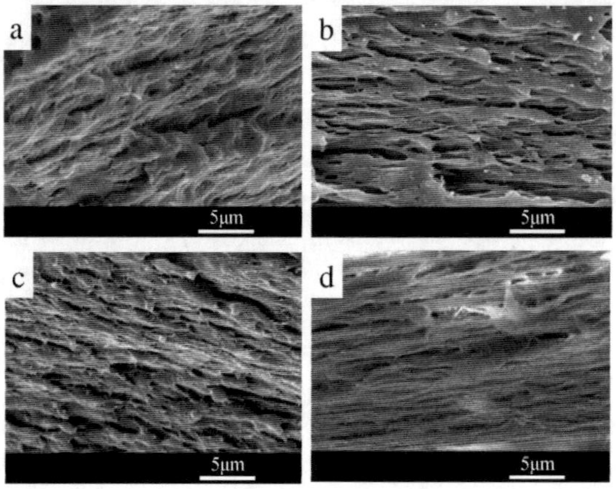

Figure 2. Impact post-fracture surface of PLA-b-PCL copolymer (30 wt. % PCL content) with and without PIF-processing characterized by SEM: (a) original sample, (b, c, d) samples via PIF-processing under the pressure of 240, 440, 540 MPa respectively.

3.2 The effect of repeating processing times on molecular weight and mechanical properties

In order to investigate on the maintenance of molecular weight and mechanical properties for PLA-b-PCL copolymer during PIF-processing, we conducted a series of repeating processing (as shown in figure 3a). The PLA-b-PCL copolymer was first cut into fragments (as shown in figure 3b) and then was reprocessed via exactly the same PIF-processing. The pieces of PLA-b-PCL were not transparent, but it showed good transparence after reprocessing as shown in figure 3c. After recycled for ten times, the copolymer is still processable at ambient temperature.

The molecular weight and mechanical properties of reprocessed samples were studied in figure 4. The results showed that with increasing reprocessing times the weight-average molecular weight (Mw) reduced slightly and polydispersity gradually broadened (see figure 4a). Mechanical properties including impact strength, tensile strength and Young's modulus are shown in figure 4(b, c, d) respectively which showed only modest and negligible decline. It indicated that PLA-b-PCL copolymer reprocessed by PIF-processing maintained good mechanical performances and the plastic was fully recyclable. Similar outstanding material has been denominated as baroplastic owing to its excellent processibility under low temperature [29].

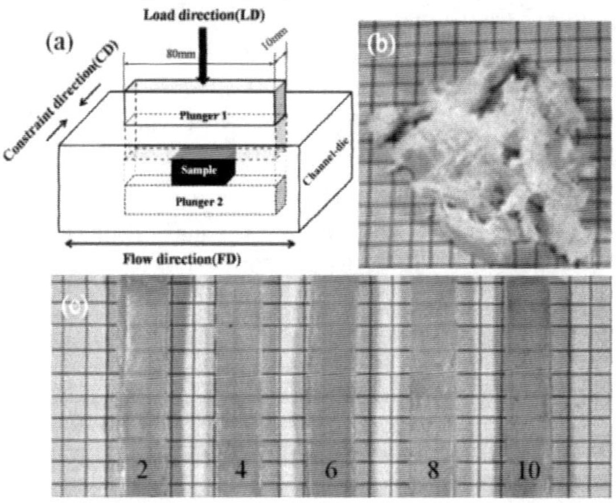

Figure 3. Reprocessed copolymer samples (30 wt. % PCL content). (a) schematic diagram of PIF-processing, (b) sample pieces after being cut, (c) the samples reprocessed for 2, 4, 6, 8, 10 times respectively. (The samples were all processed at 30 °C under 440 MPa for 10 min.)

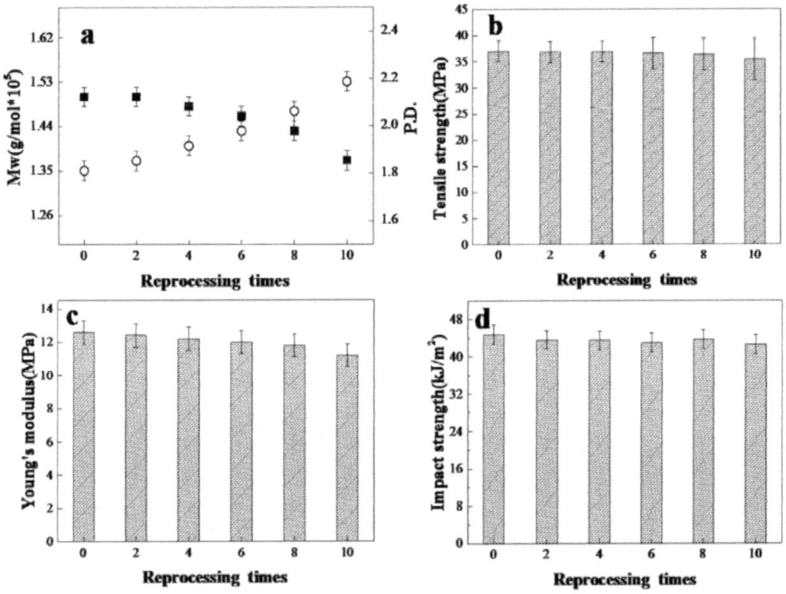

Figure 4. The molecular weight, polydispersity and mechanical properties of PLA-b-PCL (30 wt. % PCL content) copolymer after cutting and reprocessing. (a) molecular weight and polydispersity, (b) tensile strength, (c) Young's modulus and (d) impact strength versus PIF-processing times. (The samples were all processed at 30 °C under 440 MPa for 10 min.)

4. Conclusions

The impact toughness of PLA-b-PCL was enhanced notably via a physical pressure-induced-flow (PIF) processing at relatively low temperature of ~30°C (much below T_m). Simultaneously, the tensile strength and Young's Modulus were not sacrificed but also got a raise, which might expand the application of PLA-b-PCL. Furthermore, the PLA-b-PCL copolymer samples were able to reprocess for ten times without significant decrease of molecular weight or mechanical properties. Such unique behavior may reduce the cost of manufacturing of polymers, avoid serious degradable problems in processing and promote recycling of materials.

Acknowledgement

The authors acknowledge the financial support from the National Natural Science Foundation of China (No. 21404023) and the Fundamental Research Funds for the Central Universities (2232018D3-16), Shanghai Science & Technology Commission (No. 16DZ1121404), Innovation Fund for National Commercial Aircraft Manufacturing Engineering Technology Research Center (COMAC-SFGS-2016-33310).

References

[1] Zhang S, Feng X, Zhu S, et al. 2013 *Mater. Lett.* **98**:238-41.
[2] Ma H, Hu J, Ma PX. 2010 *Adv. Funct. Mater.* **20**:2833-41.
[3] Meng B, Tao J, Deng J, et al. 2011 *Mater. Lett.* **65**:729-32.
[4] Pillin I, Montrelay N, Grohens Y. 2006 *Polymer.* **47**:4676-82.
[5] Piorkowska E, Kulinski Z, Galeski A, Masirek R. 2006 *Polymer.* **47**:7178-88.
[6] Jang T-S, Lee E-J, Kim H-E, Koh Y-H. 2012 *Mater. Lett.* **72**:157-9.
[7] Li X, Kanjwal MA, Stephansen K, Chronakis IS. 2012 *Mater. Lett.* **75**:189-91.
[8] Kim WJ, Basavaraja C, Thinh PX, Huh DS. 2013 *Mater. Lett.* **90**:14-8.
[9] Qian J, Xu M, Suo A, et al. 2013 *Mater. Lett.* **93**:72-6.

[10] Song CX, Feng XD. 1984 *Macromolecules*. **17**:2764-7.
[11] Huang M-H, Li S, Vert M. 2004 *Polymer*. **45**:8675-81.
[12] Darensbourg DJ, Karroonnirun O. 2010 *Macromolecules*. **43**:8880-6.
[13] Wang L, Ma W, Gross RA, McCarthy SP. 1998 *Polym. Degrad. Stab*. **59**:161-8.
[14] Cohn D, Salomon AH. 2005 *Biomaterials*. **26**:2297-2305.
[15] Lipik VT, Kong JF, Chattopadhyay S, et al. 2010 *Acta Biomater*. **6**:4261-70.
[16] Kouya T, Tada S-i, Minbu H, et al. 2013 *Mater. Lett*. **95**:103-6.
[17] Takayama T, Todo M, Tsuji H. J. Mech. Behav. Biomed. 2011 *Mater*. **4**:255-60.
[18] Luyt, A. S., and S. Gasmi. 2017 *J. Polym Environ*. 1-14.
[19] Phromma, Warangkhana. R. Magaraphan. 2017 *J. Polym Environ*. 1-13.
[20] Birnkrant MJ, Li CY, Natarajan LV, et al. 2007 *Nano Lett*. **7**:3128-33.
[21] García C, María J, Boronat T, et al. 2018 *Polyme*r. **10**(3). Doi: 10.3390/polym10010003
[22] Song Z, Huang X. Lu X, et al. 2018 *J. Appl. Polym. Sci.* **135**(31). Doi: 10.1002/app.46536
[23] Lezak E, Bartczak Z, Galeski A. 2006 *Polymer*. **47**:8562-74.
[24] Zhang S, Zhu S, Feng X, et al. 2013 *RSC Advances*. **3**:6879-87.
[25] Taniguchi I, Lovell NG. 2012 *Macromolecules*. **45**:7420-8.
[26] Huan Q, Zhu S, Ma Y, et al. 2013 *Polymer*. **54**:1177-83.
[27] Ritchie RO. *Nat*. 2011 *Mater*. **10**:817-22.
[28] Feng X, Zhang S, Zhu S, et al. 2013 *RSC Advances*. **3**: 11738-44.
[29] Gonzalez-Leon JA, Acar MH, Ryu SW, et al. 2003 *Nature*. **426**:424-8.

IOP Conf. Series: Materials Science and Engineering **436** (2018) 012007 doi:10.1088/1757-899X/436/1/012007

Carbon Fiber Reinforced Composite Materials for Self-supporting Subway Train Cab

Ming Gong[1,2,*], Shou-guang Sun[1] and Qiang Li[1]

[1] School of Mechanical, Electronic and Control Engineering, Beijing Jiaotong University, Beijing 100044, China;
[2]CRRC Institute, Beijing 100070, China

*ssg1684@163.com

Abstract. In order to meet the requirement of lightweight, the design of the subway self-supporting composite subway cab is highly desired. Here we designed and developed carbon fiber reinforced composite sheets with high strength and moduli to replace the aluminum alloy used in the traditional subway cab. The tensile, compression, bending, interlaminar shear properties, and high speed impact performance of composite sheets were investigated. The results showed that the composite sheets had good mechanical properties comparable to traditional aluminum sheets, while reducing the weight of the subway cab by more than 30%. This composite is expected to be a good promising material to make the new generation of subway cab.

1. Introduction

In recent years, the requirements of low-carbon and environmental protection have been urgently needed with the rapid development of the rail transit industry. Vehicle weight reduction is imminent to achieve the goals. New materials, new designs, and new manufacturing technologies are the main ways to reduce the weight of rail trains. The carbon fiber reinforced composite material has been recognized to be a promising material to reduce the weight of various transportations because of it's superior properties such as light in weight, high strength, high temperature resistance, impact resistance, corrosion resistance, creep resistance, noise reduction, designability, high fatigue strength and low notch sensitivity[1].

Composite materials have been used for accessorial and non-structural components such as floor, door, luggage rack, window frame of rail trains in France, Switzerland, the United Kingdom, Japan, South Korea and other countries for a long time[2-8]. Recently composite materials are also used for manufacturing main load bearing parts for high speed trains such as the driver's cab, the upper apex and the head of the car[9-13]. With the rapid development of trains in China, more and more composite materials have also been used to build train parts such as front nose, air deflector and framework of the driver's cab. However, most applications of the composite materials are still limited to non-load bearing structural parts in rail trains, while the main load bearing structural parts, especially the car body and skeleton structure of driver's cab are mainly made from aluminum alloy, due to the strength and stiffness of traditional composite materials cannot meet the design standards of the car body. This greatly limits the wide applications of composite materials in the rail trains.

Here we designed and developed composite sheets with high strength and moduli for subway train cab. The traditional aluminium alloy metal skeleton was replaced by Composites integrally molded reinforcing skeleton to manufacture a full-composite subway train cab was fabricated to replace the aluminium alloy based one with comparable mechanical performance and significantly reduced weight by 30%. Here we would like to report the preparation and mechanical evaluation of these composites.

2. Materials and methods

2.1. Preparation of composite sheets for mechanical properties test
The raw material of 3k carbon fiber prepreg, carbon fiber unidirectional prepreg, glass fiber prepreg, aramid prepreg were prepared and cut into the desired shape. Then, the prepreg was laid according to the preset layer structure of 3 mm composite sheet (3K twill / 0° carbon fiber / ± 45° carbon fiber / aramid plain weave / 0° glass fiber / 90° glass fiber / aramid plain weave * 2 / 90° glass fiber / 0° glass fiber / aramid plain weave / ± 45° carbon fiber / 0° carbon fiber) and 2 mm composite sheet (4° carbon fiber / -45° carbon fiber / 0° glass fiber / 90° glass fiber / aramid plain weave t * 2 / 90° glass fiber / 0° glass fiber / -45° carbon fiber / 4 ° carbon fiber). The prepreg piece should be aligned around, less air bubbles, smooth, and without residue of release paper and PE film During the laying process. The prepreg piece was put onto the corresponding position of the tempered glass scribbled with release agent. Then put them into the hot press tank to get the composite sheet.

The process conditions of hot press tank molding process was referenced Figure 1. Briefly, the temperature of hot press tank was raised from room temperature to 80 °C and kept for 30 minutes. Then, the temperature was raised 100 °C and kept for 30 minutes. And then the temperature was raised 130 °C and kept for 90 minutes and dropped to 60 °C at last. The pressure was maintained at 0.3 MPa during the temperature was between 80°C and 130 °C.

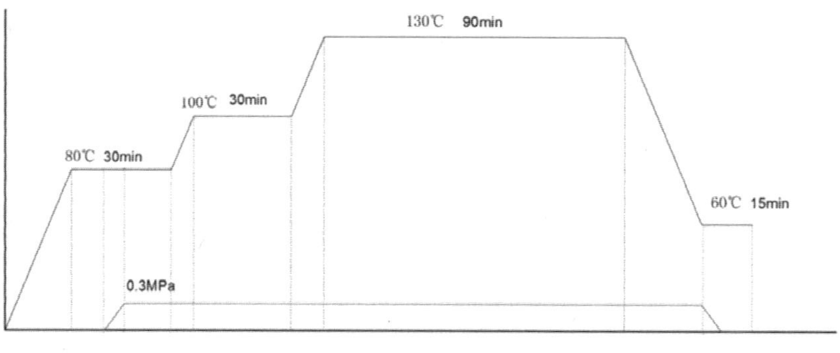

Figure 1. Process conditions of hot press tank molding process

2.2. Preparation of sandwich structure composite sheets for high speed impact performance test

2.2.1. Preparation of inner and outer skin of sandwich structure composite sheets. The prepreg were prepared and cut into the desired shape of 1100 mm * 800 mm. Then preform of inner and outer skin is constructed by the structure of 3.32 mm outer skin (3K / aramid plain weave *3 / ± 45° glass fiber / ± 45° carbon fiber / ± 45° carbon fiber / ± 45°glass fiber / aramid plain weave * 3) and 2.34mm inner skin (± 45° glass fiber / ± 45° carbon fiber / aramid plain weave * 3 / ± 45° carbon fiber / ± 45°glass fiber). And then The inner and outer skins were made according to the hot pressing process at the temperature of 130 °C, the pressure of 0.6 MPa and holding time > 1 h.

2.2.2 Preparation of sandwich structure composite sheets. 59 mm polymethacrylimide (PMI)close-cell foam board with 75% porosity was cut into the predetermined size by a woodworking saw. The inner and outer skins and the foam were bonded by an epoxy film. And then the sandwich structure

IOP Conf. Series: Materials Science and Engineering **436** (2018) 012007 doi:10.1088/1757-899X/436/1/012007

composite sheet was compounded according to the hot pressing process at the temperature of 130 °C , the pressure of 0.3 MPa and holding time > 1 h. The composite sheet was cut into the test sample with the size of 1000*700mm by the computer numerical control (CNC).

2.3. Resin content and porosity test of composite sheets
The Resin content and porosity of composite sheets were evaluated according to the standard of American Society for Testing Materials (ASTM) D3171-2015. The data from 3 specimens were averaged.

2.4. Mechanical properties test of composite sheets
Composite sheets with a thickness of 2 mm and 3 mm were prepared and cut into the test standard size. The mechanical properties of the composite sheets were evaluated by tension, compression, bending, interlayer shear tests using an Instron 5985 universal testing machine according to the standards of ASTM. The data from 5 specimens were averaged.

2.4.1. Tensile test of composite sheets. The tensile mechanical properties of the composite sheets were measured by 0°and 90° tensile test according to the standards of ASTM D3039-2014. The sections of the test specimen were rectangle, ang the sizes of them were 250 mm * 25 mm * 3 mm and 250 mm * 25 mm * 2 mm. 2 mm thick reinforcing pieces were adhibited to both ends of the specimen. The test was at a tensile rate of 2 mm / min using extensometer to measure the longitudinal strain.

2.4.2. Compressive test of composite sheets. The compressive mechanical properties of the composite sheets were measured by 0° and 90° compressive test according to the standard of ASTM D3410-2016. The sections of the test specimen were rectangle, ang the sizes of them were 150 mm * 10 mm * 3 mm and 150 mm * 10 mm * 2 mm. The test was at a compressive rate of 1.5 mm / min using a strain gauge adhibited in the middle of the specimen to measure the vertical strain.

2.4.3. Bending test of composite sheets. The sending mechanical properties of the composite sheets were measured by 0°and 90° compressive test according to the standard of ASTM D7264-2015. The sections of the test specimen were rectangle, ang the sizes of them were 120 mm * 15 mm * 3 mm and 120 mm * 15 mm * 2 mm. The distance between the fulcrum of test specimen was 97mm for 0° compressive test and 65mm for 90° compressive test and the test was at a compressive rate of 1 mm / min.

2.4.4. Interlayer shear mechanical properties test of composite sheets. The interlayer shear mechanical properties of the composite sheets were measured according to the standard of ASTM D2344-2013. The sections of the test specimen were rectangle, ang the sizes of them were 20 mm * 6 mm * 3 mm. The distance between the fulcrum of test specimen was 12 mm and the test was at a compressive rate of 1 mm / min.

2.5. Mechanical properties test of open-cell composite sheets
The mechanical properties of open-cell composite sheets were measured by compressive test according to the standard of ASTM D6484-2014. The sections of the test specimen were rectangle, ang the sizes of them were 300 mm * 36 mm * 3 mm and 300 mm * 36 mm * 2 mm and the diameter of the hole was 5.6 mm. The test was at a compressive rate of 1 mm / min.

2.6. High speed impact performance test of sandwich structure composite sheets
High speed impact performance of composite sheets was measured by a bird collision experiment equipment (F-17 NZ-II) at 20 °C and 45% relative humidity according to the standards of UIC CODE 651-2002. The sizes of the test specimen was 1000 mm * 700 mm * 60 mm. A 1.0 kg aluminum bullet

impactor impacted the specimen at a speed of 304.93 km/h (Vmax + 160 km/h, Vmax was the maximum design speed) during the test.

3. Results and discussion

3.1. Resin content and porosity test of composite sheets

The resin content of 2 mm sandwich structure composite sheets was 35.56 ± 0.3%, and the specimen has a Low porosity of 0.52 ± 0.16% to keep the composite sheets have an excellent performance.

3.2. Mechanical properties test of composite sheets

3.2.1. Tensile mechanical properties test of composite sheets. The 0° tensile strength of the 2 mm and 3 mm composite sheets were 356.37 ± 8.10 MPa and 626.24 ± 20.5 MPa and the 90° tensile strength of the 2mm and 3mm composite sheets were 525.32 ± 20.20 MPa and 376.49 ± 13.62 MPa respectively (Table 1.). The 0° tensile modulus of the 2 mm and 3 mm composite sheets were 32.65 ± 0.36 GPa and 43.51 ± 0.37 GPa and the 90° tensile modulus of the 2 mm and 3 mm composite sheets were 32.53 ± 0.40 GPa and 30.30 ± 0.67 GPa respectively (Table 1.). They were all comparable to aluminium alloy sheets (6082) used in the subway cab now according to the standards of TB/T3260-2011.

Table 1. Tensile mechanical properties of composite sheets

Thickness	Tensile strength (MPa)		Tensile modulus (GPa)	
	0°	90°	0°	90°
2 mm	356.37 ± 8.10	525.32 ± 20.20	32.65 ± 0.36	32.53 ± 0.40
3 mm	626.24 ± 20.5	376.49 ± 13.62	43.51 ± 0.37	30.30 ± 0.67

3.2.2 Compressive mechanical properties test of composite sheets. The 0° compressive strength of the 2 mm and 3 mm composite sheets were 288.74 ± 12.56 MPa and 238.01 ± 18.16 MPa and the 90° compressive strength of the 2mm and 3mm composite sheets were 264.63±15.56 MPa and 222.77 ± 16.94 MPa respectively (Table 2.). The 0° compressive modulus of the 2 mm and 3 mm compressive sheets were 61.31 ± 0.90 GPa and 78.38 ± 4.48 GPa and the 90° compressive modulus of the 2 mm and 3 mm composite sheets were 66.91 ± 2.94 GPa and 55.11 ± 1.04 GPa respectively (Table 2.). They were all comparable to aluminium alloy sheets (6082) used in the subway cab now according to the standards of TB/T3260-2011.

Table 2. Compressive mechanical properties of composite sheets

Thickness	Compressive strength (MPa)		Compressive modulus (GPa)	
	0°	90°	0°	90°
2 mm	288.74 ± 12.56	264.63 ± 15.56	61.31 ± 0.90	66.91 ± 2.94
3 mm	238.01 ± 18.16	222.77 ± 16.94	78.38 ± 4.48	55.11 ± 1.04

3.2.3 Bending mechanical properties test of composite sheets. The 0° bending strength of the 2 mm and 3 mm composite sheets were 687.73 ± 13.08 MPa and 865.17 ± 58.29 MPa and the 90° bending strength of the 2 mm and 3 mm composite sheets were 776.86 ± 57.33 MPa and 538.98 ± 43.19 MPa respectively (Table 3.). The 0° bending modulus of the 2 mm and 3 mm compressive sheets were 50.89 ± 2.36 GPa and 62.61 ± 1.35 GPa and the 90° bending modulus of the 2mm and 3mm composite sheets were 40.06 ± 0.89 GPa and 39.79 ± 1.40 GPa, respectively (Table 3.). They were comparable to than atluminum alloy sheets (6082) used in the subway cab now according to the standards of TB/T3260-2011.

3.2.4 Interlayer shear mechanical properties test of composite sheets. The interlayer shear strength of the 3mm composite sheets were 38.56 ± 2.47 MPa according to the results of interlayer shear mechanical properties tests.

IOP Conf. Series: Materials Science and Engineering **436** (2018) 012007 doi:10.1088/1757-899X/436/1/012007

Table 3. Bending mechanical properties of composite sheets

Thickness	Bending strength (MPa)		Bending modulus (GPa)	
	0°	90°	0°	90°
2 mm	687.73 ± 13.08	776.86 ± 57.33	50.89 ± 2.36	40.06 + 0.89
3 mm	865.17 ± 58.29	538.98 ± 43.19	62.61 ± 1.35	39.79 ± 1.40

3.3. Mechanical properties test of open-cell composite sheets

The tensile strength of the 2mm and 3mm open-cell composite sheets were 369.88 ± 25.01 MPa and 461.13 ± 15.60 MPa and the compressive strength of the 2 mm and 3 mm open-cell composite sheets were 219.55 ± 15.65 MPa and 203.83 ± 11.75 MPa respectively (Table 4.). They were all comparable to atluminum alloy sheets (6082) used in the subway cab now according to the standards of TB/T3260-2011.

Table 4. Mechanical properties of open-cell composite sheets

Thickness	Tensile Strength (MPa)	Compressive strength (MPa)
2 mm	369.88 ± 25.01	219.55 ± 15.65
3 mm	461.13 ± 15.60	203.83 ± 11.75

3.4. High speed impact performance test of sandwich structure composite sheets

The sandwich structure composite material meet the standard requirement of the subway according to the standards of UIC CODE 651-2002.

4. Conclusion

We designed and developed composite materials with high strength and moduli for subway train cab. The composite materials meet the standard requirements of the subway and can be used to manufacture a full-composite subway train cab to reduce the weight of the vehicle by 30% under the premise of ensuring the same performance.

References

[1] Wennberg D and Stichel S 2014 Multi-functional design of a composite high-speed train body structure *Struct Multidiscip O* **50** 475-488
[2] Wennberg D, Stichel S and Wennhage P 2012 Optimisation of sandwich panels for the load carrying structure of high-speed rail vehicles *Int J Aerosp Light Struct* **2** 19-40
[3] Harte A, McNamara J and Roddy I 2004 A multilevel approach to the optimisation of a composite light rail vehicle bodyshell *Compos Struct* **63** 447-453
[4] Hudson C, Carruthers J and Robinson A 2010 Multiple objective optimisation of composite sandwich structures for rail vehicle floor panels *Compos Struct* **92** 2077-82
[5] Kim J, Lee S and Shin K 2007 Manufacturing and structural safety evaluation of a composite train carbody *Compos Struct* **78** 468-476
[6] Belingardi G, Cavtorta M and Duella R 2003 Material characterization of a composite-foam sandwich for the front structure of a high speed train *Compos Struct* **61** 13–25
[7] Lee Y, Kim J, Lee H, Kil K and Park B 2000 A study the application of 3D sandwich composite structures to the double-deck light train carbody *J Korean Soc Railway* **3** 92–99
[8] Wu Z, Zeng J and Liu J 2011 Development of High-speed Train and Its Composites *Materials Review* **25** 108-114
[9] Belingardi G, Cavatorta M and Duella R 2003 Material characterization of a composite–foam sandwich for the front structure of a high speed train *Compos Struct* **61** 13-25
[10] Zinno A, Fusco E, Prota A and Manfredi G 2010 Multiscale approach for the design of composite sandwich structures for train application *Compos Struct* **92** 2208-19
[11] Kim J, Kim N, and Han S 2005 Optimal stiffness design of composite laminates for a train carbody by an expert system and enumeration method *Compos Struct* **68** 147-156
[12] Kim J and Jeong J 2006 Natural frequency evaluation of a composite train car body with length of 23 m *Compos Sci Techn* **66** 2272-83
[13] Kim J and Chung S 2007 A study on the low-velocity impact response of laminates for composite railway bodyshells *Compos Struct* **77** 484-492

Poly (Terephthaloyl Diglyceride-co-1, 2-cyclohexanediacyl Diglyceride) - A Biodegradable and Biocompatible Functional Polymer for Biomedical Applications

YangfenXie[1], Lijie Sun[2], Xiaoping Bi[3,4], Yifan Guo[2], Shuo Chen[2], Xianqun Fan[3,4], Fengling Qing[1] and Zhengwei You[2,*]

[1] State Key Laboratory for Modification of Chemical Fibers and Polymer Materials, College of Chemistry, Chemical Engineering and Biotechnology, Donghua University, 2999 North Renmin Road, Shanghai 201620, China
[2] State Key Laboratory for Modification of Chemical Fibers and Polymer Materials, College of Materials Science and Engineering, Donghua University, 2999 North Renmin Road, Shanghai 201620, China
[3] Department of Ophthalmology, Ninth People's Hospital, Shanghai JiaoTong University School of Medicine, No.639 Zhizaoju Road, Shanghai, 200011, China.
[4] Shanghai Key Laboratory of Orbital Diseases and Ocular Oncology, Shanghai, 200011, China.

*Corresponding author: zyou@dhu.edu.cn; Tel: (+86 02167874279)

Abstract. Biodegradable polyesters have been widely used in biomedical engineering. Introduction of functional groups to polyesters will expand their applications. Herein, we developed a functionalizable copolyester with free hydroxyl groups (-OH) - poly (terephthaloyl diglyceride-co-1,2-cyclohexanediacyl diglyceride) (PTeCD). PTeCD was readily synthesized by acid-induced epoxide ring-opening polymerization in one step. Its structure was characterized by NMR and FTIR. PTeCD exhibited a good cytocompatibility and biodegradability. We expect PTeCD will be a useful polymer for biomedical applications.
Key words: biodegradable, aliphatic-aromatic copolyester, functional

1. Introduction

Biodegradable polymers have been used in a wide range of emerging biomedical applications including temporary prostheses, tissue engineering scaffolds, and drug delivery vehicles [1]. Aliphatic polyesters such as poly(ε-caprolactone) (PCL) and poly(lactic acid) (PLA) represent one of the most important classes of biodegradable polymeric biomaterials [1-3]. Comparing with enzymatically degradable polymers, aliphatic polyesters are usually hydrolysable, and thus, are preferred for *in vivo* applications due to the minimal site-to-site and patient-to-patient variation of hydrolysis [4]. To meet the diverse requirements of biomedical applications, the properties of polymeric biomaterials are expected to be modulated in a wide range. Introduction of aromatic moieties to the polymer structure is an efficient way to modulate the properties such as thermal and mechanical performance [5]. However, compared to aliphatic polyesters, aromatic polyesters have been less studied as biodegradable biomaterials partly due to their limited biodegradability. Furthermore, most synthetic polymeric biomaterials including polyesters are generally biologically inert and lack of reactive groups for modifications, which limit the modulation of their properties and consequent applications [6-7]. Introduction of functional groups could be one way to address aforementioned drawbacks. Functional groups such as hydroxyls can directly tune the physical, chemical and biological properties of polymers. Furthermore, modification of these functional groups provides access to diverse polymers with a wide range of properties [4, 8-9]. However, chemical synthesis of functional polyesters is still a challenge in polymer science, because it usually involves a complex multistep process with a low overall yield [10-12]. Recently, we established a new simple and versatile method, acid-induced epoxide ring-opening polymerization, which could efficiently synthesize polyesters with pendent hydroxyl groups from a variety of diacids and diepxides in one step [13]. It is appliable on both aromatic and aliphatic substrates [14-15]. Accordingly we set out to use this new developed method to synthesize functional aliphatic/aromatic copolyesters, which would be useful biodegradable polymers with tunable properties. Here, we report one example, poly(terephthaloyl diglyceride-*co*-1,2-cyclohexanediacyl diglyceride) (PTeCD) (**Figure 1**). We designed this functional polyester according to following consideration. PTeCD contains alternating aliphatic and aromatic repeating units. PTeCD are expected to be biodegradable. In addition, PTeCD has extensive hydroxyl groups, which would improve the hydrophobic nature of aromatic structure and make PTeCD to have suitable hydrophilicity for biomedical applications. More importantly, hydroxyl groups are robust for transformations including facile biofunctionalization with various biomolecules such as peptide and gene. Thus, PTeCD would be a versatile intermediate for a series of functionalized biomaterials.

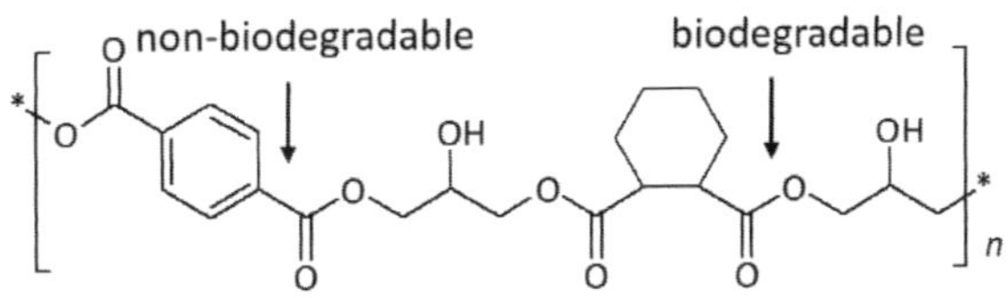

Figure 1. Structure of PTeCD

2. Experimental Section

2.1 Materials and reagents

Diglycidyl 1,2-cyclohexanedicarboxylate (J&K, 95%), terephthalatic acid (Acros Organics, 98%), tetrabutylammonium bromide (Acros, ≥99%), anhydrous solvents N,N-dimethylformamide (DMF), and 1,6-hexyldiisocyanate (Sigma, ≥99%) were used without further purification. Deuterated dimethyl sulfoxide (DMSO-d_6) for NMR analysis was purchased from Cambridge Isotope Laboratories, Inc. All other solvents were purchased from Sinopharm Chemical Reagent Co., Ltd.

2.2 Synthesis and characterization of PTeCD

Diglycidyl 1,2-cyclohexanedicarboxylate, an equal molar amount of terephthalatic acid, and 0.6 mol% tetrabutylammonium bromide were mixed and dissolved in anhydrous DMF into a three-neck round-bottom flask in a glove box filled with nitrogen. The flask was transferred out of the glove box in a sealed state, and connected to nitrogen. Then, the contents in flask were stirred under a nitrogen atmosphere at 100 °C for 26 h. The reaction mixture was purified via precipitation in ethyl ether for three times, and vacuum-dried at ambient temperature for 24 h to yield PTeCD.

The molecular weight and distribution of PTeCD were determined via Waters gel permeation chromatography (GPC) using a differential refractive index detector and a Brookhaven multi-angle light scattering detector. The measurement was performed at 40 °C using DMF (HPLC grade) as the eluent. ^{1}H nuclear magnetic resonance (NMR) spectrum was obtained in DMSO-d_6 using a Bruker 400 NMR (400 MHz) spectrometer at 25 °C. Fourier transformed infrared (FTIR) was carried out using a Thermo Nicolet 6700 spectrometer with a total of 32 scans from 500 to 4000 cm^{-1}.

2.3 Evaluation of in vitro biocompatibility of PTeCD

The wells of a tissue culture-treated polystyrene (TCPS) 24-well plate were coated with 80 μl of 1 g/l 2,2,2-trifluoethanol solution of PTeCD that filtered through a 0.2 μm sterilization filter. The plate was dried under vacuum overnight to evaporate the residual solvent in polymer coating. Both the PTeCD-coated plate and TCPS plate were sterilized under UV light for 30 min, then washed with 1 ml phosphate buffered saline (Lonza) at least three times and with 1 ml culture medium once with gentle shaking (20 rpm). Each well was seeded with 50,000 nonimmortalized rat primary osteoblasts obtained from neonatal (2–3 days old) Spratue-Dawley rats, and isolated by sequential trypsin-collagenase digestion of calvaria. The cells were cultured in Dulbecco's phosphate-buffered saline (DPBS) supplemented with Dulbecco's Modified Eagle Medium (DMEM) and fetal calf serum at 37 °C with 5% CO_2. The culture medium was exchanged every 2 days. The cell adhesion rate was used as a quantitative evaluation of osteoblast adhesion onto PTeCD coating and TCPS plates (n = 5). Osteoblasts were cultured for 3 and 24 h on the polymer-coated plates. At each time point, the adhered osteoblasts were enzymatically (0.25% trypsin–0.1% EDTA; Gibco) detached and counted using hemocytometer. The cell adhesion rate was expressed as a percentage of the initial number of seeded cells (2×10^4 cells per well).

2.4 In vitro degradation of PTeCD
Disk-shaped sample of PTeCD (n = 4) with dimensions of 5 mm × 2 mm (diameter × thickness) was evenly cut into 4 pieces of fan-shaped specimens. Every specimen was weighted and incubated at 37 °C in 1 ml DPBS solution with a pH of 7.2-7.4. At the predetermined period of times, buffer solution was exchanged, and the samples washed with deionized water and lyophilized at least 24 h in sequence. Mass remained of sample was determined by the following formula (where M_0 and M_f are the initial mass and dry-weight at each time point, respectively):

$$\text{Mass remained (\%)} = \frac{M_f}{M_0} \times 100\,\% \tag{1}$$

3. Results and Discussion

3.1 Polymer synthesis and characterization

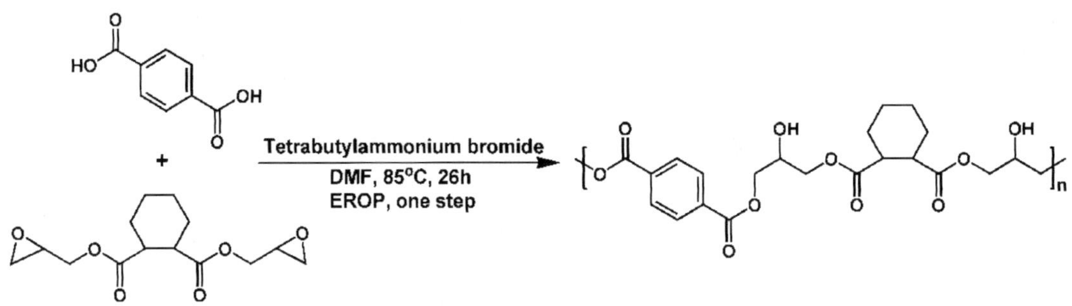

Figure 2. Synthesis route of PTeCD.

PTeCD (M_n = 9.2 kD, PDI = 2.1) was prepared by acid-induced epoxide ring-opening polymerization between commercially available diglycidyl 1,2-cyclohexanedicarboxylate and terephthalatic acid in the presence of a catalytic amount of tetrabutylammonium bromide (Bu_4NBr).(**Figure 2**)

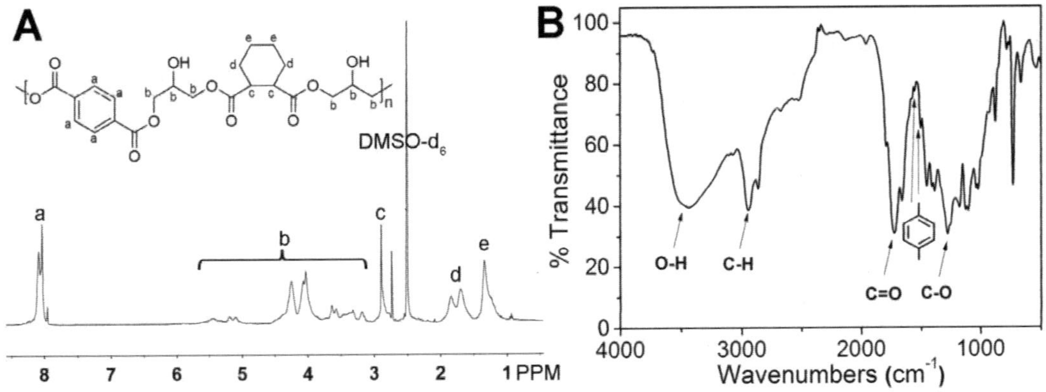

Figure 3. Characterization of PTeCD. (A) 1H NMR (DMSO-d_6) spectrum. (B) FTIR spectrum.

The structure of PTeCD was characterized by ^{1}H NMR and FTIR. In the ^{1}H NMR spectrum (**Figure 3A**), the signal marked 'a' at chemical shift δ values of 8.06 ppm was assigned to the CH protons of benzene ring. The three signals marked 'c, d and e' at chemical shift δ values of 2.88, 1.69-1.85 and 1.33 ppm were ascribed to the three different protons of the cyclohexane, respectively. The signals of CH and CH_2 protons in the glyceryl moiety marked 'b' appeared at δ 3.19-5.45 ppm. To further confirm the presence of the characteristic functional groups in PTeCD, FTIR spectrum was recorded. As shown in **Figure 3B**, a broad and intense peak at ~3489 cm^{-1} was in accordance with the stretching vibration of the O-H bonds. The intense peak around 1731 cm^{-1} was assigned to C=O stretching vibration, confirming the formation of ester bonds. Skeletal vibration of benzene ring displayed signals at 1506 cm^{-1} and 1577 cm^{-1}, as well as the C-O and C-H stretching vibration at 1278 cm^{-1} and 2947 cm^{-1}, respectively.

3.2 Evaluation of in vitro biocompatibility of PTeCD.

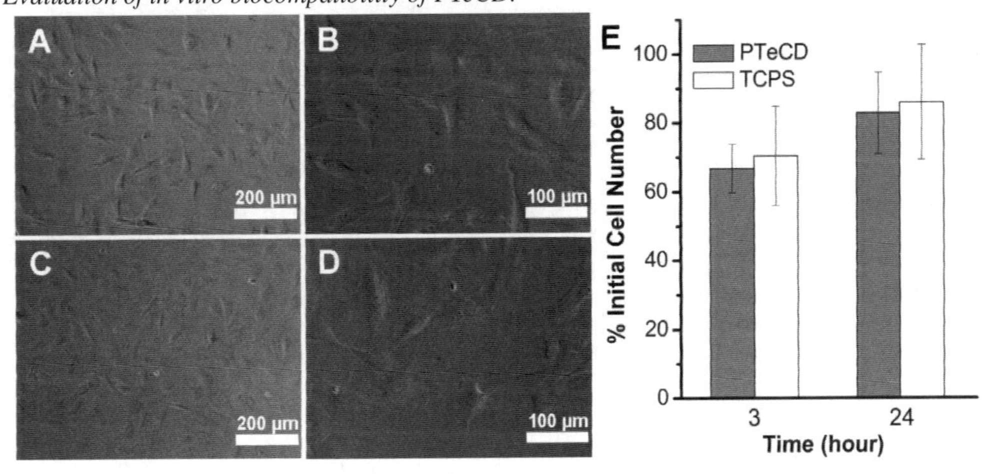

Figure 4. Representative phase contrast photomicrographs of osteoblasts at 24 hours after seeding in PTeCD-coated wells (A, B) and TCPS wells (C, D). (E) Attached cell numbers of osteoblasts in PTeCD-coated wells and TCPS wells 3 h and 24 h after seeding.

The *in vitro* biocompatibility of PTeCD was evaluated by culturing rat primary osteoblasts on PTeCD-coated 24-well plates using pristine TCPS as a control. The osteoblasts cells showed similar polygonal morphology on PTeCD (**Figures 4A** and **4B**) and TCPS (**Figures 4C** and **4D**). Moreover, the adhesion of osteoblasts on PTeCD was investigated by quantitative counting adhered cells on the surface at 3 and 24 hours after seeding. As shown in **Figure 4E**, the cell adhesion rate on PTeCD coating was at a similar level to that on TCPS at each time point, reaching 66.8 ± 7.0% and 82.8 ± 11.9% at 3 and 24 h, respectively. This indicated that PTeCD supported osteoblast adhesion. These results reveal PTeCD had a good cytocompatibility.

3.3. In vitro degradation of PTeCD.

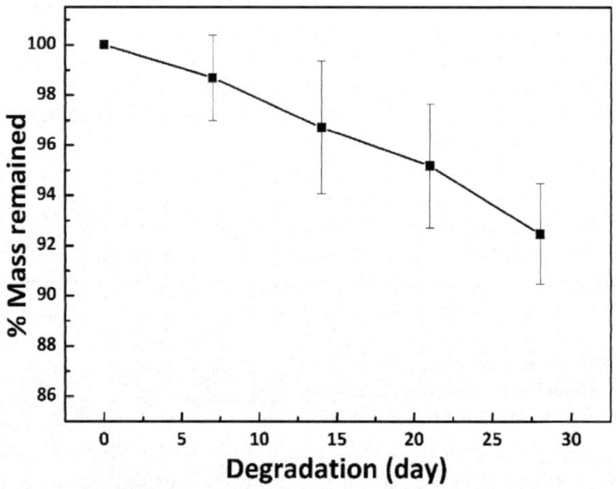

Figure 5. *In vitro* degradation of PTeCD in DPBS.

We evaluated the in vitro degradation of PTeCD in DPBS. PTeCD showed a steady degradation with a remained mass of 92.5 ± 2.0% at day 28 (**Figure 5**). This confirmed the degradability of PTeCD.

4. Conclusions

An aliphatic-aromatic copolyester (PTeCD) with free hydroxyl groups was designed and successfully synthesized in one step from commercially available materials. PTeCD exhibited good cytocompatibility and degradability. The modification of free hydroxyl groups of PTeCD can modulate its physical, mechanical, and biological properties. We expect PTeCD will be a useful polymer for biomedical applications.

Acknowledgement

The authors thank the National Natural Science Foundation of China (21574019, 81320108010, 81470662), the Natural Science Foundation of Shanghai (18ZR1401900), and the Science and Technology Commission of Shanghai (17DZ2260100).

References

[1] L. Wang, S. Venkatraman and L. Kleiner, *Journal of controlled release : official journal of the Controlled Release Society*, 2004, **99**, 207-216.
[2] J. W. Leenslag and A. J. Pennings, *Macromol. Chem. Phys.*, 1987, **188**, 1809-1814.
[3] D. Garlotta, *J. Polym. Environ*, 2001, **2**, 63-84.
[4] M. Vert, *Biomacromolecules*, 2005, **6**, 538-546.
[5] I. Pomerantseva, N. Krebs, A. Hart, C. M. Neville, A. Y. Huang and C. A. Sundback, *J Biomed Mater Res A*, 2009, **91**, 1038-1047.
[6] L. S. Nair and C. T. Laurencin, *Prog. Polym. Sci.*, 2007, **32**, 762-798.
[7] Z. Ma, Z. Mao and C. Gao, *Colloids Surf B Biointerfaces*, 2007, **60**, 137-157.

[8] D. S. Benoit, M. P. Schwartz, A. R. Durney and K. S. Anseth, *Nat. Mater.*, 2008, **7**, 816-823.
[9] C. K. Williams, *Chem. Soc. Rev.*, 2007, **36**, 1573-1580.
[10] D. E. Noga, T. A. Petrie, A. Kumar, M. Weck, A. J. García and D. M. Collard, *Biomacromolecules*, 2008, **9**, 2056-2062.
[11] R. J. Pounder and A. P. Dove, *Polym. Chem.*, 2010, **1**, 260-271.
[12] S. Ponsart, J. Coudane and M. Vert, *Biomacromolecules*, 2000, **1**, 275-281.
[13] Z. You, H. Cao, J. Gao, P. H. Shin, B. W. Day and Y. Wang, *Biomaterials*, 2010, **31**, 3129-3138.
[14] Z. W. You, X. P. Bi and Y. D. Wang, *Macromol. Biosci.*, 2012, **12**, 822-829.
[15] Z. You and Y. Wang, *Adv. Funct. Mater.*, 2012, **22**, 2812-2820.

IOP Conf. Series: Materials Science and Engineering **436** (2018) 012009 doi:10.1088/1757-899X/436/1/012009

Bearing Fault Diagnosis Method Based on Hilbert Envelope Demodulation Analysis

Nan WANG* **and Xia LIU**

Xi'an High-Tech Institute, Xi'an, 710025, P. R. China

*wangnan_2018edu@163.com

Abstract. A digital envelope analysis method based on Hilbert transform is proposed to solve the problems of bearing fault signals such as non-stationary, faintness and non-linearity. This method firstly compares the power spectrums between the fault signal and the normal signal. In order to improve the signal-to-noise ratio, a frequency band with a large difference between the two power spectrums is selected. Secondly, the frequency band is used to remove the high frequency carrier and retain the low frequency part which contains fault information only. In order to avoid the wrapping effect in the circular convolution, the tail of the frequency band is filled with zero to double its length. Then the inverse Fourier transform and the Fourier transform are applied to the transformed signal. Finally, the fault diagnosis is made for its envelope spectrum. This method is applied to the bearing fault signal of case Western Reserve University in the United States. The results show that this method can diagnose the fault type and fault location well and has high sensitivity.

1. Introduction

Rolling bearing is one of the most commonly used parts in all kinds of rotating machinery and one of the vulnerable parts of rotating machinery. According to statistics, 30% of the fault of rotating machinery is caused by bearing failure, so it has significant effects on the working condition of the machine greatly. Bearing defects may cause severe vibration and noise of the machine, and even the equipment damage. Therefore, it is very necessary for us to carry out research in condition detection and fault diagnosis for bearings [1].

At present, scholars at home and abroad have done a lot of research on the diagnosis method of bearing vibration analysis. In the literature [2], a bearing fault diagnosis method based on variational modal decomposition and envelope demodulation is proposed. Literature [3] has proposed an improved fault diagnosis method for rolling bearing with envelope spectrum based on empirical mode decomposition and spectral kurtosis. Literature [4] combines wavelet packet with approximate entropy to diagnose cylindrical roller bearing. Literature [5] combines harmonic wavelet packets transform with information entropy and is successfully used in the diagnosis of three bearing conditions. These methods for fault diagnosis provides the necessary means, its basic principle is based on the actual situation to choose a high frequency vibration mode as the research object, through the center frequency is equal to the natural frequency of the band pass filter to the analysis of vibration signal is extracted, and then through the envelope detector detection, remove the high frequency attenuation vibration frequency components, get only contains the fault feature information of low frequency envelope signal, then carries on the envelope analysis [6]. The method of envelope demodulation analysis based on Hilbert transform is applied to bearing fault diagnosis. From the fault mechanism of rolling bearing, the failure characteristic frequency can be clearly seen by analyzing the signal with the method of Fourier transform. The fault location is determined by filtering the vibration signal and detecting the envelope signal by envelope demodulation.

2. Vibration signal characteristics of rolling bearings

The vibration of rolling bearings can be caused by external vibration sources or by the structural characteristics and defects of the bearings themselves. The vibration signal has rich frequency components and wide frequency band, including bearing failure characteristic frequency and vibration characteristic frequency caused by machining and assembling error. This frequency band is susceptible to the influence of other parts and structures in the machine tool, and the energy of the characteristic frequency component information reflecting the damage type of fault impact is small at the initial stage of failure, and the signal-to-noise ratio is low. Medium frequency band (1~20 kHz), includes the natural vibration frequency of bearing outer ring caused by the surface damage of bearing original. By analyzing the vibration signal in this frequency band, the damage fault of bearing can be diagnosed. Due to bearing damage caused by shock induced bearing high frequency vibration generally in more than 20 kHz, if measuring sensor resonance frequency is higher, bearing fault can also be diagnosed by analysis of this band [7].

The calculation formula of bearing failure characteristic frequency is obtained by analyzing the relative motion relation between bearing elements. Let the rotation frequency of the bearing be f, the bearing pitch is D, the diameter of the rolling body is d, the contact Angle is φ, the number of stroller is n, then we can get:

$$BPFI_{Inner\ ring\ fault} = \frac{nf_r}{2}\left(1+\frac{d}{D}COS\varphi\right) \tag{1}$$

$$BPFO_{Outer\ ring\ fault} = \frac{nf_r}{2}\left(1-\frac{d}{D}COS\varphi\right) \tag{2}$$

$$BSF_{Rolling\ body\ failure} = \frac{Df_r}{2d}\left[1-\left(\frac{d}{D}COS\varphi\right)^2\right] \tag{3}$$

$$FTF_{Cage\ failure} = \frac{f_r}{2}\left(1-\frac{d}{D}COS\varphi\right) \tag{4}$$

Generally speaking, the failure characteristic frequency of rolling bearing is less than 1 kHz, which is one of the important characteristic information of rolling bearing failure. When the bearing fails, the fault characteristic frequency spectral line will appear in its vibration spectrum in theory, but in practice, these spectral lines cannot always appear in the theoretical value accurately. In general, the edge frequency band of the characteristic frequency spectrum line and its harmonic center line is modulated by the rotation frequency of the shaft. These phenomena are mainly caused by the error of bearing installation and other conditions, but generally will not affect the diagnosis of rolling bearing.

3. Data Envelopment Analysis

3.1. Hilbert demodulation analysis

When there is a fault in the rolling bearing, each time the rolling body passes through the fault position, an impulse of impact will be generated, which will cause the high frequency natural vibration of the bearing. With the vibration frequency as the carrier, the passing frequency of the impulse is modulated by the modulated wave. Make the bearing vibration waveform eventually showed the characteristics of amplitude modulation and frequency modulation, using band-pass filter to remove low frequency interference, only by modulation of the high frequency component, through demodulation technique to fault information separated from complex amplitude modulation signal.

The Hilbert transform of the vibration signal and the original signal are used to form a set of analytic signals, which are envelope signals.

Its Hilbert transform is defined as:

IOP Conf. Series: Materials Science and Engineering **436** (2018) 012009 doi:10.1088/1757-899X/436/1/012009

$$\hat{x}(t) = \frac{1}{\pi}\int_{-\infty}^{+\infty} \frac{x(\tau)}{t-\tau} d\tau = \frac{1}{\pi}\int_{-\infty}^{+\infty} \frac{x(t-\tau)}{\tau} d\tau$$
$$= x(t) * \frac{1}{\pi t} \tag{5}$$

The analytic signal is composed of Hilbert transform and original signal:

$$z(t) = x(t) + j\hat{x}(t) = a(t)e^{jf(t)} \tag{6}$$

$$a(t) = \sqrt{x^2(t) + \hat{x}^2(t)} \tag{7}$$

$$\phi(t) = \arctan\left[\frac{\hat{x}(t)}{x(t)}\right] \tag{8}$$

It can be seen from the formula that the envelope of the analytic signal is the envelope of the original vibration signal with certain transformation in amplitude. Fault information can be obtained by envelope demodulation of this analytic signal. In order to improve the signal-to-noise ratio of the modulated signal, the demodulation analysis is carried out for a certain frequency band with a large difference between the fault signal and the normal signal.

(1) The band pass filter is used to extract the frequency range that needs to be modulated and move to 0 frequency, as shown in fig.1.

As can be seen from figure 1, the amplitude of the analytic signal after passing through the band pass filter has not changed.

(2) From the Fourier transform of the analytic signal, we can see

$$Z(j\Omega) = X(j\Omega) + j\hat{X}(j\Omega) = X(j\Omega) + jH(j\Omega)X(j\Omega) \tag{9}$$

So

$$Z(j\Omega) = \begin{cases} 2X(j\Omega) & \Omega > 0 \\ 0 & \Omega < 0 \end{cases} \tag{10}$$

The amplitude of the analytical signal in the positive frequency part is twice that of the original signal, and the analytical signal at the negative frequency is 0. So we can move the low frequency part that contains the fault information to the left of the sampling frequency, the inverse Fourier transform to the time domain. Or as *c* and *d* in figure 2 multiply their amplitude by 2, and add 0 to double their length.

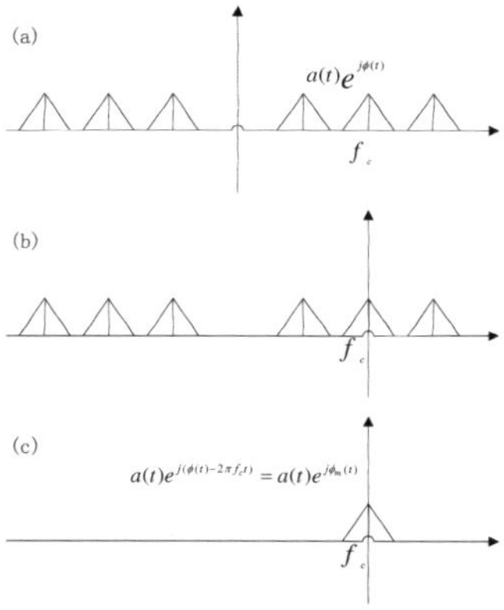

Figure 1. Extraction of a frequency band by a zoom processor.

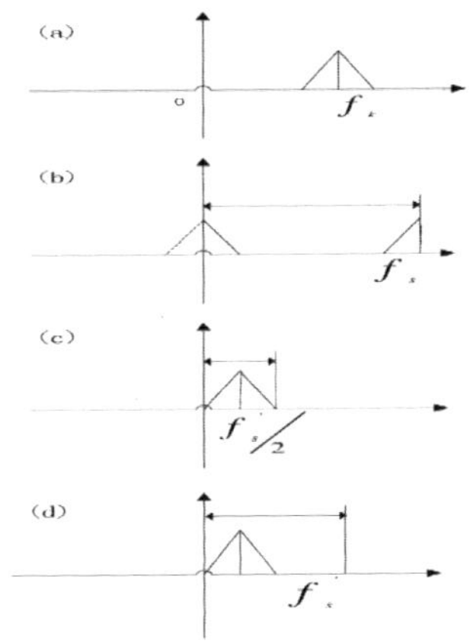

Figure 2. Block shift procedure for selection of frequency band for inverse transformation.

(3) Inverse Fourier transforms the shifted frequency to the time domain. At this time, the time domain signal contains amplitude modulation information caused by failure.

$$x(t)=\frac{1}{2\pi}\int_{-\infty}^{+\infty}X(f)e^{j2\pi ft}df \tag{11}$$

(4) Take the envelope of the signal and make the Fourier transform.

$$X(f)=\int_{-\infty}^{+\infty}x(t)e^{-j2\pi ft}dt \tag{12}$$

By analyzing the envelope spectrum of the signal, it can be clearly seen that the envelope of the signal at some frequency points presents a mutation, which is the result of amplitude modulation.

Envelope analysis is one of the most effective methods in fault diagnosis. It can clearly indicate the location and severity of the failure. The filtering, translation and transformation of vibration signal are also called envelope demodulation. Envelope mediation can mediate fault related signals from high frequency modulation signals, so as to avoid confusion with other low frequency interference, and has high diagnostic reliability and sensitivity.

3.2. Bearing fault diagnosis process based on Hilbert transform

The bearing fault diagnosis method based on Hilbert transform is one of the most effective and direct methods.

Based on the selected frequency signal is a series of Fourier transform and inverse Fourier transform, used in the final frequency spectrum of the signal envelope diagram has a corresponding frequency point mutation compared with different types of fault characteristic frequency, in order to complete fault diagnosis.

The method steps are shown in figure 3:

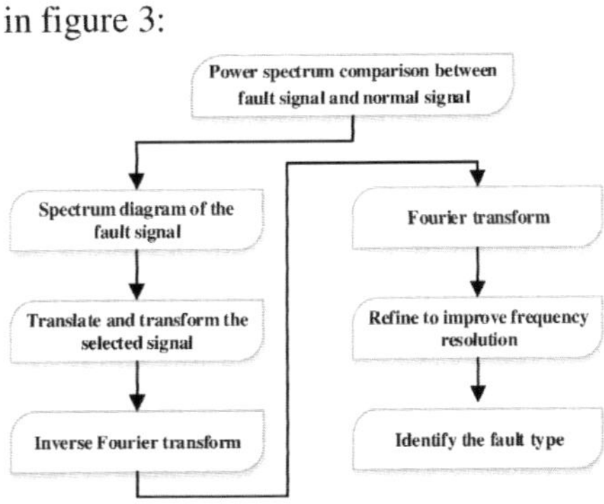

Figure 3. The flow chart of bearing fault diagnosis based on envelope analysis.

4. Bearing fault diagnosis example

4.1. Introduction to the experimental platform
According to the vibration signal measured by case Western Rcserve University [8, 9], the feasibility of this method is verified. The experimental table is shown in figure 4:

Figure 4. Bearing fault simulation test bench of CWRU.

The test bearing is a deep groove ball bearing of SKF 6205-2rs type. The specific specifications are shown in table 1 (unit: mm). The fault of bearing is the pitting failure caused by electric spark on the bearing surface. The vibration signal of the bearing is collected by the vibration acceleration sensor mounted on the drive end of the motor and above the bearing seat of the fan end.

Table 1. SKF 6205-2rs deep groove ball bearing parameter specification

Inner ring diameter	outer ring diameter	Rolling diameter	Pitch diameter	Number of rolls
25.0012	51.9989	7.94	39.0398	9

The bearing fault data used in this paper include the vibration signals of outer ring, inner ring, rolling body fault and normal state [10-12]. These signals are collected at the motor drive end with a sampling frequency of 12000Hz and a motor speed of 1750r/min. The outer ring due to the influence of load and show the fault status is different, so the external circle respectively at 3 o'clock and 6 o'clock and 12 o'clock direction orthogonal direction, namely the fault processing center direction and the opposite direction.

4.2. Vibration signal envelope analysis results
By comparing the power spectrum of fault signal and normal signal, it can be found that there is a big difference between them. This provides convenience for the extraction of the frequency band. In this paper, the power spectrum analysis of the inner ring signal and the normal signal is carried out. The comparison results are shown in figure 5:

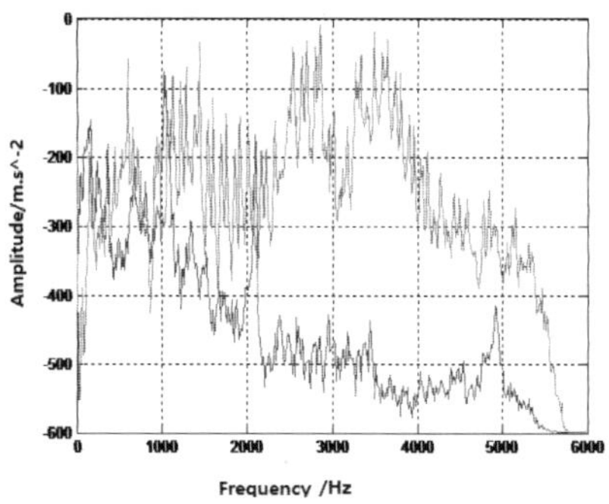

Figure 5. Inner ring signal is compared with the normal signal power spectrum.

According to the bearing parameters in table 1 specification, combining the calculation formula of the failure frequency of the mentioned in the article for the outer ring fault characteristic frequency of

104.56 Hz, the inner ring of the fault characteristic frequency of 157.94 Hz, fault characteristic frequency of 137.48 Hz, roller cage fault characteristic frequency of 11.63 Hz, in addition to turn the frequency of 29.3 Hz. The analysis process of the third part of the paper is used to further process the signal. The envelope analysis results are shown in figure 6:

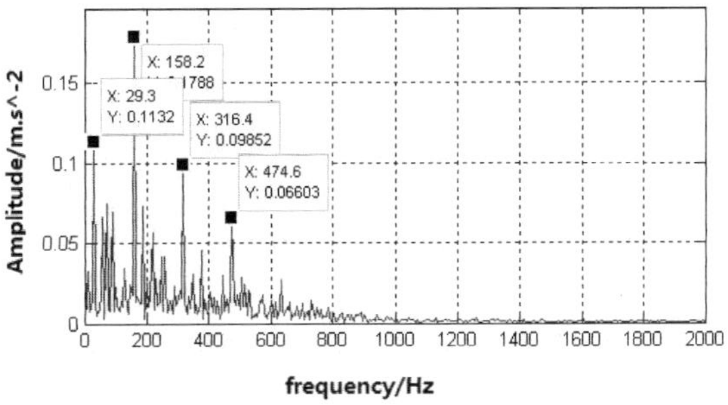

Figure 6. Envelope analysis result of inner ring signals.

It is can be clearly observed from the figure, 29.3 Hz, 158.2 Hz, 316.4 Hz, and 158.2 Hz four obvious peaks, 29.3 Hz frequency corresponds to the transform, and 158.2 Hz, 316.4 Hz and 316.4 Hz is very close to the inner ring fault frequency of 157.9 Hz and its frequency doubling, so it can conclude that the bearing inner ring failed.

The same method is used to analyze the outer ring signal and the rolling body signal respectively. The analysis results are shown in figure 7 and 8:

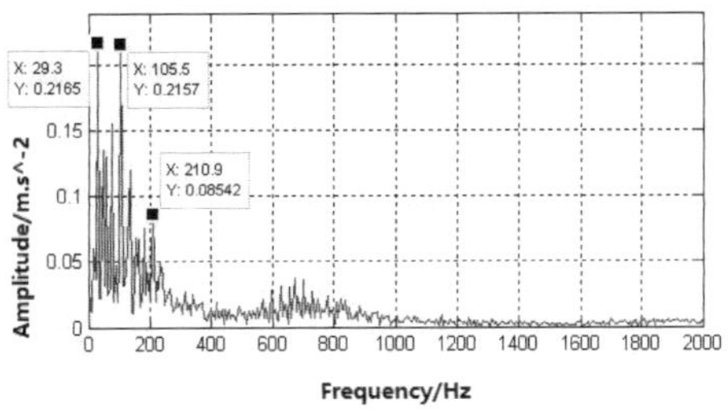

Figure 7. Envelope analysis result of outer ring signals

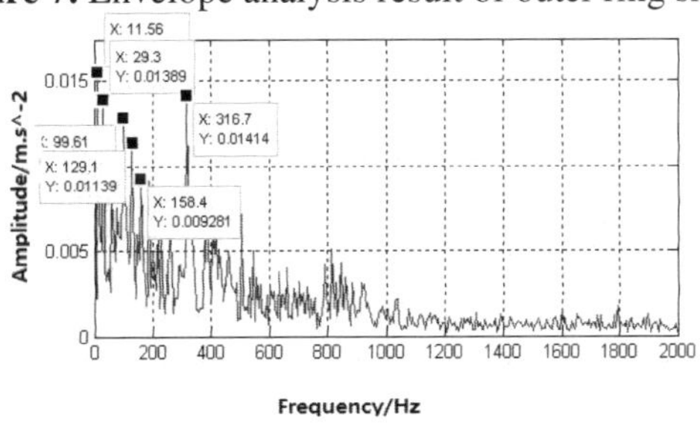

Figure 8. Envelope analysis result of rolling element signals

Figure 7 is the spectral envelope line bearing outer ring fault feature component, can be seen in the figure at 29.3 Hz, 105.5 Hz and 210.9 Hz obvious peaks, of which 105.5 Hz and 210.9 Hz very close to the outer ring fault characteristic frequency of 104.56 Hz and twice the frequency of 209.12 Hz, it can be concluded that the outer ring failure frequency. However, there is a certain gap between the failure frequency and the theoretical value, which is mainly caused by skidding, friction and inaccurate measurement of bearing parameters. 29.3Hz corresponds to the frequency of revolution, because the outer ring rotates on the axis, and it can be found that there are equally spaced modulation spectral lines on both sides of 105.5Hz, which is the modulation effect of frequency conversion.

Figure 8 is the envelope spectrum line of the fault feature components of the bearing rolling body.

It can be clearly seen from the figure that 29.3Hz, 11.56Hz, 99.61Hz, 129.1Hz and 158.4Hz are respectively. While 29.3Hz corresponds to the frequency conversion, 11.56Hz corresponds to the fault characteristic frequency of the cage. For 99.61Hz, 129.1Hz and 158.4Hz, the difference between them is 29.3Hz. At the same time, it can also be found that the fault characteristic frequency of 129.1Hz is close to 137.48 Hz of the rolling body, so it can be known that there is a fault in the rolling body. Compared with the former two kinds of fault, fault features of rolling frequency differ with the theoretical value is bigger, this is because the rolling element distribution, outer ring and cage, is strongly influenced by them, own roller rotation phenomenon at the same time, so the information contained in the vibration signal is much more complicated.

5. Conclusion

(1) Since the vibration signal of rolling bearing is periodic, the periodic vibration pulse caused by bearing fault causes high frequency natural vibration of bearing. So, the fault information can be attached to the high frequency vibration signal, and the vibration signal has the characteristics of frequency modulation and amplitude modulation due to the modulation effect.
(2) Digital envelope demodulation method used in this article can automatically calculate the maximum frequency in the power spectrum contrast difference, improve the signal-to-noise ratio in the analysis and facilitate subsequent fault information diagnosis research.
(3) To avoid the surround effect in the cyclic convolution, we has carried on the zero padding to move after the frequency operation, thus making the transform length doubled, and sampling frequency to twice, so the frequency resolution is ensured by this way.

References

[1] B. Li, M. Yuen Chow, Y. Tipsuwan, J.C. Hung, Neural-network-based motor rolling bearing fault diagnosis, IEEE Trans. Ind. Electron. 47 (5) (2000) 1060–1069.
[2] TANG Guiji, WANG Xiaolong. Parameter Optimized Variational Mode Decomposition Method with Application to Incipient Fault Diagnosis of Rolling Bearing[J].Journal of Xi'an JiaoTong University, 2015, 49 (5): 73-80.
[3] Ding JianMing, Lin Jianhui, Ren Yu,et al.I.C. REAL-TIME DIAGNOSIS OF BEARING FAULTS BASED ON HARMONIC WAVELET ENERGY ENTROPY [J]. Journal of Mechanical Strengh, 2011, 33(4): 483-487.
[4] Li Xuejun, He Nengsheng, He Kuanfang. Cylindrical Roller Bearing Diagnosis Based on Wavelet Packet Approximate Entropy and Support Vector Machines [J]. Journal of Vibration, Measurement&Diagnosis, 2011, 35(6): 1031-1036.
[5] Cheng Fabin, Tang Baoping, Liu Wenyi. A Method to Suppress Cross- terms of Wigner- Ville Distribution and Its Application in Fault Diagnosis [J]. China Mechanical Engineering, 2008, 19 (14) 1727-1731.
[6] Cai Yanping, Li Aihua, Shi Linsuo,et al.Roller bearing fault detection using improved envelope spectrum analysis based on EMD and spectrum kurtosis[J]. Journal of Vibration and Shock, 2011, 30(2): 167-172.
[7] C. Wang, G. Too, Rotating machine fault detection based on HOS and artificial neural networks, J. Intell. Manuf. 13 (2002) 283–293.
[8] R.B. Randall, J. Antoni, Rolling element bearing diagnostics—a tutorial Feb., Mech. Syst. Signal Process. 25 (2) (2011) 485–520.
[9] El-Thalji, E. Jantunen, A descriptive model of wear evolution in rolling bearings, Eng. Fail. Anal. 45 (2014) 204–224.
[10] L. Guo-lu, Z. Zhi-qiang, W. Hai-dou, X. Bin-shi, P. Zhong-yu, Z. Li-na, Acoustic emission monitoring and failure mechanism analysis of rolling contact fatigue for Fe-based alloy coating May, Tribol. Int. 61 (2013) 129–137.
[11] M. Elforjani, D. Mba, Accelerated natural fault diagnosis in slow speed bearings with acoustic emission Jan., Eng. Fract. Mech. 77 (1) (2010) 112–127.
[12] Y. He, X. Zhang, M.I. Friswell, Defect diagnosis for rolling element bearings using acoustic emission, J. Vib. Acoust. 131 (6) (2009).

Effect of Freezing and Thawing on Chloride Ion Erosion of Fiber Concrete

B D Zhao[1,a] and Z Y Sun[1,b]

[1]School of Architectural and Civil Engineering, Shenyang University, Shenyang, 110044, China

[a]zhaobaidong@163.com; [b]zhiyousun@163.com

Abstract: the frost resistance of ordinary concrete is poor, and concrete is often destroyed in the engineering practice at low temperature. This test takes 1% polypropylene fiber concrete as an example to explore the effect of different mix ratio and different freezing thawing times on the diffusion of chloride ion. After repeated freezing and thawing, concrete members will be subjected to structural damage that can not be ignored. As the number of freeze-thaw increases, concrete shows a gradual increase in the mass loss rate and a relative decrease in relative dynamic elastic modulus. Whether or not the freeze-thaw cycle is applied, the chloride penetration of concrete increases with the increase of water cement ratio. In the freezing-thawing environment, the chloride diffusion of concrete increases rapidly with the increase of freeze-thaw times, finally, the least square method is used to fit the second law of Fick, and the mathematical solution of the new equation of chloride diffusion in concrete is obtained.

1. Introduction

In cold regions or winter climatic conditions, concrete undergoes freeze-thaw damage, causing internal micro-cracks and surface flaking; chloride ions in the environment invade the concrete and cause steel corrosion. For example, chloride ions in seawater in coastal areas and chloride ions introduced when concrete bridge decks are sprinkled with ice salts. The combination of the two ultimately leads to the destruction of the concrete structure, which is one of the reasons why many reinforced concrete structures fail prematurely before they have reached the design life.

Xie Youjun[1] et al. used the fast freezing method to explore the frost resistance of ordinary concrete in different solutions. The results showed that different concretes had different frost resistance in different solutions, and the frost resistance of concrete was affected by the chemical corrosion resistance of the concrete salt solution. Yang Wenwu[2] et al. conducted a comparison of the experimental and evaluation methods for frost resistance and atmospheric ion diffusion of concrete in the marine environment. Based on an analysis of the comparison of concrete frost resistance and ambient ion diffusion experiments and evaluation methods. The suitable experimental method for studying the frost resistance and diffusion of marine concrete was found.By simulating the severe climate environment of winter and night temperature difference during winter construction in severe cold area, Li Jianguo[3] et al. adopted the chloride diffusion coefficient as a token of the loss degree of concrete after the frost damage, and analyzed the structure development after the concrete early experience of freezing thawing cycle with different water cement ratio, thus enriching the increasingly perfect winter. Sun Congtao and Niu Yutao[4] et al. studied the chloride ion diffusion performance of fly ash concrete under ocean underwater zone, ocean tide zone, ocean atmosphere zone and freeze-thaw conditions, established a chloride ion diffusion model in fly ash concrete, and predicted the life of concrete structures under chloride ion erosion. The anti-salting performance and anti-salt freezing quantitative design of fly ash concrete were studied by experiments. Zhao Baidong and Xu Tiancheng[5], et al., by testing the chloride ion flux of self-compacting concrete with various fiber contents, it is concluded that the six-hour total electric flux of self-compacting concrete with various reference strengths is related to its fiber content. With the increase of fiber content, the six-hour total electric flux decreased first and then increased.

Based on the present study, this experiment mainly explores the effect of different water to cement ratio on the diffusion of chlorine ions in different freeze-thaw cycles, and then using the least square method to fit the second law of Fick, the mathematical solution of the new equation of chloride ion diffusion in concrete is obtained.

2. Test materials and methods

2.1. *Test materials and model preparation*

The water used in this experiment is drinking water for the residents of Shenyang city. The cement is a common portland cement marked 325#, with a density of 3.12g/cm3; the gravel size is 8-17mm, and its performance index is detailed in Table 1. The sand is made of medium coarse river sand in Shenyang. The fineness modulus is 2.95, and its performance is in line with the requirements of "GBT14684-2011 building sand". The fiber is made of polypropylene fiber. Polypropylene fiber has the advantages of light weight and improved durability of concrete. It is a synthetic fiber with a low elastic modulus. According to the specification[6], the concrete block made by this test is 100mm×100mm×400mm cuboid test block. Before the test block preparation, the test should check whether the materials meet the test standard, whether the cement is hydrated and whether the aggregate diameter is up to standard. The strength of the ordinary concrete is designed to be C30.

Table 1. Main performance index of crushed stone.

Density (kg/m^3)	Crushed value (%)	Absorption (%)	Sulfide and sulfate content (%)	Mud content (%)	Mud block content (%)
2600	16	0.04	0.3	0.35	0.27

2.2. *Polypropylene fiber additive*

Fiber reinforced concrete has attracted the attention of scholars and engineering circles due to its advantages such as good crack resistance, corrosion resistance, strong erosion resistance and good frost resistance. The damage of concrete is generally manifested as the emergence and development of cracks. Adding fiber in concrete can effectively inhibit the diffusion of cracks and improve the long-term performance of concrete structures. Many scholars at home and abroad have been relatively mature in the research of polypropylene fiber reinforced concrete. It is considered that the concrete with 0.5%-1.2% fiber parameters has better frost resistance. The design of this mix ratio is based on 1% fiber content. The fiber is made of polypropylene fiber. Polypropylene fiber has the advantages of light weight and improved durability of concrete. It is a synthetic fiber with a low elastic modulus. The basic parameters of polypropylene fiber are shown in Table 2.

Table 2. Performance parameters of fiber.

Name	Diameter /μm	Length /mm	Tensile strength /MPa	Elastic modulus /GPa	Density /(g/cm^3)
Polypropylene fiber	30	10	276	3.8	0.91

2.3. *Mix ratio design*

Many evaluation indexes of concrete are related to the water cement ratio. Whether the design of water cement ratio is reasonable or not will directly affect the quality of the test results. It should be verified in a strict sense in accordance with the steps of mix proportion design. Under the same conditions of production and standard curing, the lower the water cement ratio, the higher the strength of the specimen, the higher the density and the less the pore. When cement is hydrated, the water cement ratio has a direct effect on the porosity of hardened cement paste, and porosity affects the pore volume of concrete. The pore diameter of the concrete is expanded and connected to each other, and the reduction of the number of candidate holes leads to a reduction in the effect of relieving the heaving pressure, which causes a higher expansion pressure after freezing, which increases the volume of the pores. The quality of the aggregate mixture of $1m^3$ concrete is 2450kg, as shown in Table 3. According to the concrete mix design steps, the final design and determination of concrete water cement ratio are three kinds, namely, W/C=0.40,0.45,0.50 three grades.

Table 3. Mix ratio of fiber concrete with different water/cement ratio.

Concrete number	Water cement ratio	Water	Cement	Stone	Sand	Polypropylene fiber	Collapsing degree /mm
		Unit: kg/m^3					
W_1	0.4	172	430	909	914.5	24.5	153
W_2	0.45	175	388.9	914	947.6	24.5	165
W_3	0.5	178	356	918	973.5	24.5	180

3. Freeze-thaw test and analysis

The two freeze-thaw test methods used in the experiment are quick freezing and thawing methods. This test adopts a rapid water freeze-thaw test. The freeze-thaw temperature was -20℃-20℃, and freeze-thaw time is two hours apart for one hour. The test instruments used include JCD freezing and thawing machine, TM- II dynamic instrument, electronic scale and so on. The number of freezing and thawing cycles on the whole is 200 times, and measuring the lateral frequency fm of the test piece after every 50 intervals, and the lateral fundamental frequency is measured by the dynamic elastic measuring instrument. Before the test, the surface scum was removed and the clear water was kept, and then the weight of the specimen is weighed. After testing, adjust the direction of the test block in the original test box and continue the experiment. In appearance, the external surface of concrete is mainly manifested in the reduction of the mass and the decrease of the dynamic elastic modulus, while in the mechanical properties, the compressive strength and the flexural strength are gradually attenuated. In the process of freezing and thawing test, the performance of the block can be changed in many aspects. Therefore, the following cases occur in the test process, which can end the test: the number of freezing and thawing cycle reaches the set value; the relative dynamic elastic modulus of the specimen falls to 0.6; the quality of the specimen is lower than the initial value of 95%. The test instrument and process are shown in Figure 1 and the relative elastic modulus and mass loss rate curves of this test are shown in Figure 2 and Figure 3 respectively.

In order to ensure that the whole test block is immersed in the water, the experimental treatment of 3mm of water immersion concrete was taken before the freezing and thawing test to isolate the effect of air factors on the freezing and thawing test.

Figure a.

Figure b.

Figure c.

Figure d.

Figure 1. Freeze-Thaw test

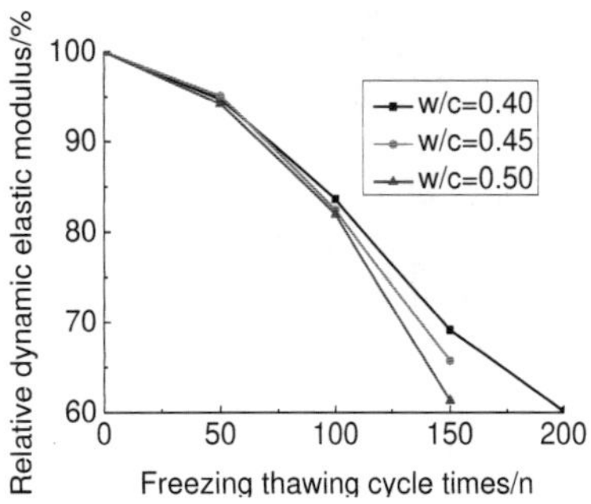

Figure 2. The relative dynamic modulus of elasticity of concrete under different mixing ratios.

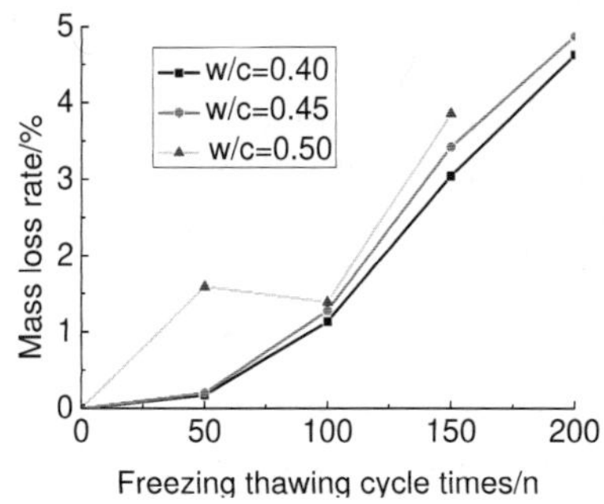

Figure 3. The quality loss rate of concrete under different mix ratios.

According to Figure 1 and Figure 2, the relative elastic modulus of concrete decreases with the increase of freeze-thaw cycles. When the freezing and thawing times are less than 100 times, the damage of concrete is not obvious, and the difference between the curves of different mix ratio is little. The possible reason is that the internal aggregate of concrete is interacted to resist the freezing and thawing. When the number of freezing and thawing times is more than 100 times, the relative dynamic elastic modulus of concrete drops faster, when the water ash is larger, the freeze-thaw resistance of concrete is relatively weak, and the relative dynamic elastic modulus of concrete with water cement ratio of 0.5 is less than 60%. The possible reason is that the number of freezing thawing times is too much, and the concrete has lost its strength.

According to Figure 1 and Figure 3, the mass loss rate has some defects as an index of freeze-thaw resistance. The quality of concrete with different mix proportioning has increased characteristics during freeze-thaw cycles. This is mainly due to a large number of micro cracks in the concrete, and the micro cracks absorb water saturated during freeze-thaw cycles. When the number of freeze-thaw cycles is low, the weight loss of concrete surface is smaller, and the quality of the specimen increases after freezing and thawing. Because of the higher density of the other test parts, the quality of the water quality in the immersed specimen is less than that of the test piece, so the quality of the specimen after the freezing and thawing cycle is heavier than that before the freeze-thaw cycle.

4. Chloride ion diffusion equation

The diffusion coefficient is the physical quantity describing the diffusion velocity. It is equal to the mass or particle number of $1m^2$ area in 1s when the concentration gradient is 1. The greater the the chloride diffusivity, the faster the diffusion rate. The chloride diffusivity (D) in concrete is an important index reflecting the durability of concrete structure to resist chloride ion. At present, the methods of detecting chloride diffusivity in concrete include natural immersion method, electromigration method and salt saturation conductivity method.

At present, most scholars study the diffusion of chloride ions based on the most commonly used chloride diffusion equation, that is, one dimensional diffusion equation derived from the second law of Fick is:

$$\frac{\partial c}{\partial t} = D\frac{\partial^2 c}{\partial x^2} \tag{1}$$

Formula: t is time, x is distance from concrete surface, d is chloride ion diffusion coefficient, c is chlorine ion concentration from concrete surface x.

The mathematical equation is:

$$c_x = \left(c_s - c_0\right)\left[1 - erf\left(\frac{X}{2\sqrt{Dt}}\right)\right] + c_0 \tag{2}$$

The chlorine ion concentration on the surface x of the concrete surface of the cx-t time, the chloride ion concentration at the cs-c0, the initial concentration of chloride ions in the concrete material, the D for the chloride ion diffusion coefficient and the erf(g) as the error function.

The equation (1) is simplified when the material is not composed of chlorine ions in the selection of concrete:

$$c_x = c_s\left[1 - erf\left(\frac{X}{2\sqrt{Dt}}\right)\right] \tag{3}$$

At present, most studies have taken chlorine ions into the steel surface as the mark of the end of the durability life of chloride ion erosion of concrete structure, and the scholar Ma Yali[7] use formula (4) to calculate the durability of the concrete structure under chloride ion erosion environment:

$$T = \frac{X^2}{4D}\left[erf^{-1}\left(1 - \frac{c_{cr}}{c_s}\right)\right] \tag{4}$$

Type: x means the thickness of reinforced concrete protective layer; cs is the critical concentration of chloride ion.

Mangat[8] et al in the chloride diffusion equation, the relationship between chloride diffusion coefficient and time is considered:

$$D = D_i t^{-m} \tag{5}$$

In the formula: Di is the effective chloride ion diffusion coefficient when the time equals 1s; m is the empirical constant.

In-band(1), the diffusion equation considering the time dependence of chloride ion diffusion coefficient is obtained:

$$\frac{\partial c}{\partial t} = D_i t^{-m} \frac{\partial^2 c}{\partial x^2} \tag{6}$$

The mathematical solution is:

$$C_f = C_s \left[1 - erfx \left(2\sqrt{\frac{D_i}{1-m} t^{1-m}} \right)^{-1} \right] \tag{7}$$

t time away from the concrete surface of the total chlorine ion concentration of ct, free chloride ion concentration of cf, combined with the concentration of chloride ions cb, then the concrete chloride binding capacity r is defined as:

$$R = \frac{c_b}{c_f} = \frac{c_t - c_f}{c_f} \tag{8}$$

The chloride ion diffusion coefficient of concrete when t0 is the D0,t time when the chloride ion diffusion coefficient of concrete is D, the relationship between chloride diffusion coefficient and time of concrete is:

$$D_t = D_0 \left(\frac{t_0}{t} \right)^m \tag{9}$$

The effect of chloride ions on concrete diffusion is accelerated by the micro-cracks inside the concrete during the use, and the equivalent diffusion coefficient De can be expressed as:

$$D_e = KD_t \tag{10}$$

In the formula: K is the deterioration effect coefficient of chloride ion diffusion property of concrete.

The time dependence of chloride ion diffusion coefficient and the new diffusion equation of chloride ions affected by structural defects are as follows:

$$\frac{\partial c_f}{\partial t} = \frac{KD_0 t_0^m}{1+R} t^{-m} \frac{\partial^2 c_f}{\partial x^2} \tag{11}$$

Using the same initial conditions and boundary conditions as the (2) type, with the aid of mangat[9], the formula (7) can be used to obtain the most basic mathematical solution of the new diffusion equation of chloride ion in concrete:

$$c_f = c_0 + \left(c_s - c_0\right) \left[1 - erfx \left(2\sqrt{\frac{KD_0 t_0^m}{\left(1+R\right)\left(1-m\right)} t^{1-m}} \right)^{-1} \right] \tag{12}$$

5. Chlorine ion concentration test

The study of the durability of reinforced concrete began in the 20s of last century. The most representative durability problem is corrosion of steel bar, and the corrosion of chlorine salt is the main cause of corrosion of steel bar, and the long-term effect will affect the safety of the structure. Therefore, it is necessary to study the chloride penetration resistance of concrete, which provides a theoretical basis for preventing chloride ion erosion.

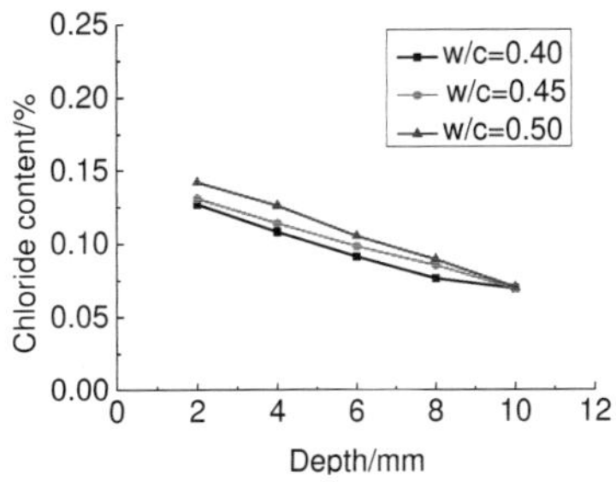

Figure 4. Chloride ion intrusion of concrete before freezing and thawing.

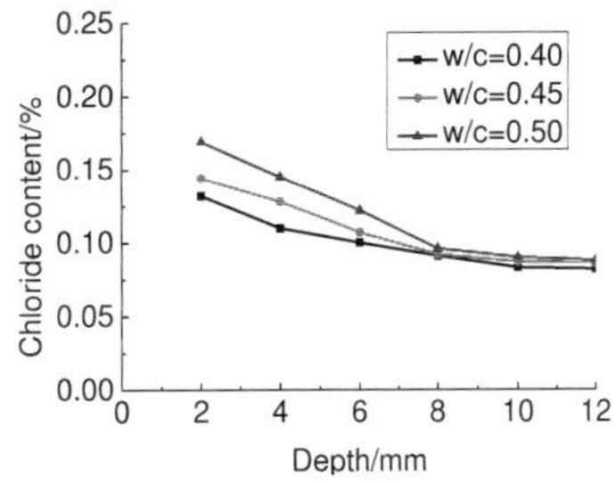

Figure 5. The penetration of chlorine ions in frozen-thawing 50 times concrete.

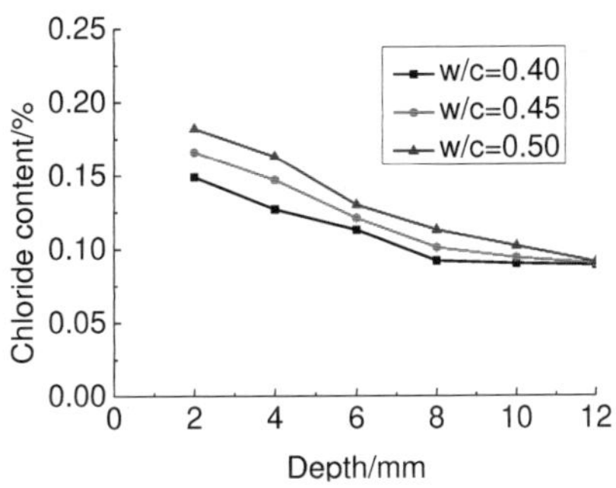

Figure 6. The penetration of chlorine ions in frozen-thawing 150 times concrete.

As shown in Figure 4, 5 and 6, the chloride penetration of concrete increases with the increase of water cement ratio, no matter whether it is freeze-thaw cycle or not. The reasons are that, on the one hand, the water cement ratio increases the internal porosity of concrete, and the diffusion of chlorine ions to the concrete is faster. On the other hand, the greater the water cement ratio, the more freeze-water content in concrete, the worse the frost resistance, and further increase the chloride ion penetration rate after the damage to the concrete surface. With the increase of the same water cement ratio, the early internal structure gradually developed, the internal porosity decreased gradually, and the chloride diffusion coefficient decreased, indicating that the increase of the pre feeding age was helpful to the formation and development of the internal structure of concrete, the development of concrete strength and the test of different water cement ratio. The greater the ratio of water to cement, the greater the diffusion coefficient of chloride ion, which shows that the concrete with large water cement ratio is relatively large, and the structure development is relatively slow because of the more water and water cement ratio. Therefore, the strength of the concrete with low water cement ratio is better than that of the concrete.

The chloride ion content of concrete after freezing and thawing cycle is obviously increased compared with the results of the chloride ion intrusion in the unthawing test block. The chloride ion mass fraction of the surface of the concrete with water cement ratio 0.45 is 0.142%, the freezing and thawing 50 times is 0.153%, and the freezing thawing increases to 0.176% when the freeze-thaw is 150 times. Freeze-thaw cycles cause cracks in concrete and accelerate chloride ion erosion. The more cycles, the more serious chloride ions attack.

6. Conclusion

(1) After repeated freezing and thawing, the concrete members will be damaged by the structure which can not be neglected, and with the increase of freezing and thawing times, the quality loss rate of concrete is improved gradually and the relative dynamic modulus of elasticity decreases gradually

with the allowable parameter range of the test.
(2) The chloride diffusion coefficient in concrete is positively correlated with the ratio of water to water cement in different specimens. The chlorine ion intrusion of concrete increases with the increase of the ratio of water to ash, which indicates that the concrete specimens with high water/cement ratio have high moisture content, so the internal porosity is relatively high, and the internal structure changes relatively slowly.
(3) In freezing-thawing environment, the diffusion of chloride ions in concrete increases rapidly with the increase of freeze-thaw times, finally, the least squares method is used to fit the second law of Fick, and the mathematical solution of the new equation of chloride ion diffusion in concrete is obtained. The mathematical solution can better reflect the actual situation of the prediction of the durability life of chloride ions.

References

[1] Youjun X,Kunlin M,Hui X and Mingxia S 2006 Study on frost resistance of concrete in different solution *J. Journal of Railway Science and Engineering.* **27** 29-34.
[2] Wenwu Y,Jueshi Q and Yubin H 2008 Experimental and evaluation method for freezing resistance and atmosphere ion diffusion of concrete in marine environment *J. Material Herald.***21** 74-77.
[3] Jianguo L,Zhiping N and Shouheng J 2013 Effect of water-cement ratio on early frost resistance of concrete *J. Low-temperature Building Technology*.**34** 17-18.
[4] Congtao S and Ditao N 2010 Study on durability and life prediction of concrete based on chloride ion erosion M.Xian: Xian University of Architecture and Technology.
[5] Tiancheng X and Baidong Z 2017 Experimental study on mechanical and durability of basalt fiber self-compacting concrete M.Shenyang: Shenyang University.
[6] National standards of the People's Republic of China.*Test method standard for concrete structure (GB/T-50152-2012) S.*Beijing:China Construction Industry Press,2012.
[7] Yali M and Ailin Z 2006 Prediction of durability life of chloride ion corrosion in concrete structures based on specified reliability index *J. Journal of Civil Engineering.***52** 36-41.
[8] Mangat P S,Limbachiya M C 1999 Effect of initial curing on chloride diffusion in concrete repair materials *J.Cement and Concrete Research* **29** 1475-85.
[9] Mangat P S,Molloy B T 1994 Prediction of long-term chloride concentration in concrete *J.Materials and Structures*.**27** 338-346.

IOP Conf. Series: Materials Science and Engineering **436** (2018) 012011 doi:10.1088/1757-899X/436/1/012011

Research on the Flame Retardant Effect of Super-porous Nitroguanidine Propellant Coated with Different Coating Contents of TiO_2

Zhang Lina[1,*], Nan Fengqiang[1] and Shen Wanwu[1]

[1] Nanjing University of Science and Technology, Nanjing 210094, China

*1124809519@qq.com

Abstract. The flame-retardant effect of coating liquid with different titanium dioxide (TiO_2) content on the coating of super-porous nitroguanidine propellants is studied by closed bomb tests and X-Ray Diffraction (XRD) tests. Using three different proportions of insensitive coating liquid to make a single-hole propellant. Taking single hole propellant 20g respectively for constant volume combustion test, and analyses the p-t, u-p curve. Then, select 30% and 35% of TiO_2 coating propellant and uncoated propellant to do constant volume combustion experiments, and to analyze the p-t curves. By analyzing the results of experiments,we draw the following two conclusions: (i) When the content of TiO_2 in the coating liquid of the single hole propellant is 35%, at the same loading density, the maximum pressure reached is the minimum and the time required is the longest. (ii) As the content of TiO_2 in the coating layer increases, the insensitivity effect of the coating layer become more obvious, and the increasing explosiveness of the propellant will increase.

1. Introduction

To reduce the burning velocity of the super-porous nitroguanidine propellant. One can choose to add different proportions insensitive materials[1]. Insensitive coating materials generally use high molecular polymers or small molecule compounds. In this paper, we utilize 100nm anatase TiO_2 as the flame retardant material. TiO_2 is a metal oxide. TiO_2 is a type of coating agents. It can prolong the ignition time and slow down the burning speed of propellants in the combustion process[2]. After adding TiO_2 as a flame retardant material to the coating liquid, a super-porous nitroguanidence propellant is insensitively coated, and the obtained coating medicine has a good sense of low-temperature[3].

TiO_2 is evenly distributed in the coating liquid, which changes the coating liquid from the original homogenous system into a heterogeneous system[4]. At the same time, because TiO_2 itself has a certain specific heat capacity, part of the heat in the combustion process of the coating will be absorbed by TiO_2, which also plays a role in flame retardant[5].

In this paper, by doing constant volume combustion experiments, analyze the p-t, u-p curve to study the flame retardant effect of coating liquids coated with different TiO_2 contents on super-porous nitroguanidine propellants.

2.Experimental

2.1. Test Materials and Instruments.

Super-porous nitroguanidine propellant, Liaoning Qingyang Special Chemical Co., Ltd.; acetone, ethanol, chemical pure, Nanjing Chemical Reagent Co., Ltd.; 100nm anatase oleophilic TiO_2, purity 99.8%, Shanghai Aladdin Biochemistry Technology Co., Ltd.; Sanwa -15 absorption tablets, Liaoning Qingyang Special Chemical Co., Ltd.

X-ray diffractometer, XP205 German Bruker company, closed explosives, Zhangzhou North Chemical Industry Co., Ltd.; coating machine, Taizhou Liming Pharmaceutical Machinery Co., Ltd.; water bath oven, AHX safe oven, Nanjing University of Science and Technology.

2.2. Experimental Methods.

25%, 30%, and 35% of the TiO_2 content of the coating liquid were made into a single-hole propellant. Doing a constant volume combustion test, and the combustion powder was taken for XRD test[6].

Select 30% and 35% TiO_2 coating propellants and uncoated propellant sing closed bomb for constant-volume combustion experiments. Exact sampling is given in Table 1.

Table 1. Constant Volume Combustion Test Sampling Table

Group number	Composition	Quality/g
(1)	Two 30% titanium dioxide coatings	38. 3984
(2)	Two 30% titanium dioxide coatings	39. 0252
(3)	Base + 30% titanium dioxide coating (mixed)	38. 2097
(4)	Two 35% titanium dioxide coatings	39. 1608
(5)	Base + 35% Titanium dioxide coating (mixed)	38. 7811
(6)	Base + 35% Titanium dioxide coating (mixed)	38. 5079

3.Data Processing and Analysis

3.1 Coating solution preparation method

(1) Placing Sanwa-15 absorption tablets in a water bath oven, and set the oven temperature to 50°C. Placing in the oven for three days, and remove the absorbent tablets after the solvent is completely evaporated.

(2) Taking a certain amount of Sanwa-15 absorption tablets, and add different amounts of TiO_2 and acetone ethanol solvent into a 2500 mL measuring cup. Using a constant digital mixer to stir at 3000 r/min and stir for about 24 h. Preparation of the coating liquid as showed in Table 2:

Table 2. Preparation of the coating liquid

Group	Sanwa-15absorbent content/g	TiO2 content/g	Solvent content/g	Solvent ration
01#	75	25 (25%)	550	1:5.5
02#	70	30 (30%)	550	1:5.5
03#	65	35 (35%)	550	1:5.5

The following “figure 1” shows the comparison of the super-porous nitroguanidine propellant coating.

IOP Conf. Series: Materials Science and Engineering **436** (2018) 012011 doi:10.1088/1757-899X/436/1/012011

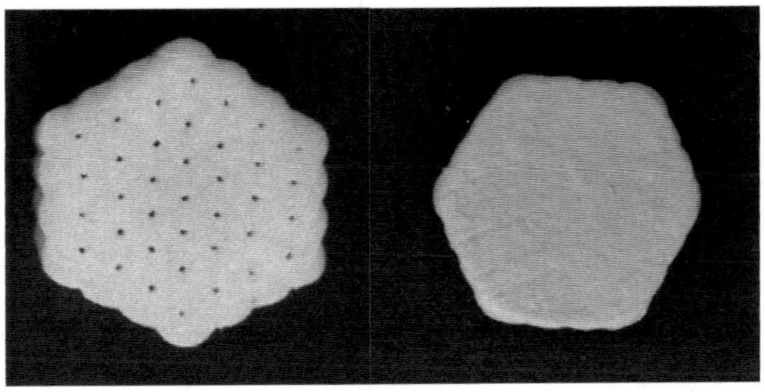

Figure 1. uncoated and coating

3.2 Effect of TiO_2 as insensitive coating material on coating.

In order to study the effect of TiO_2 content on the combustion performance of the coating. Using three different proportions of insensitive coating liquid to make a single hole propellant. Taking single hole propellant 20g respectively for constant volume combustion test, and the p-t, u-p curve are shown in the "figure 2", "figure 3".

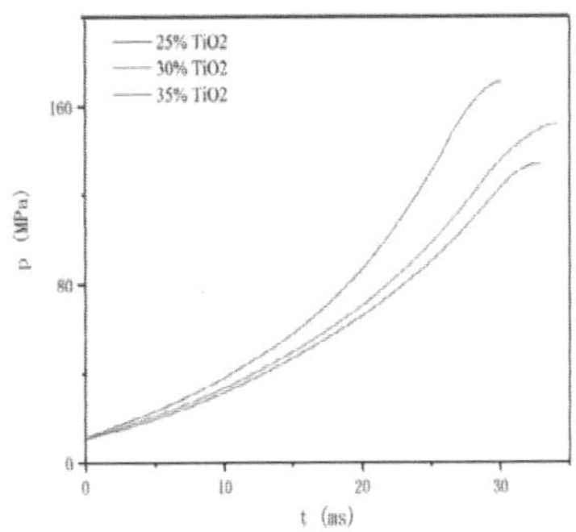

Figure 2. p-t curves

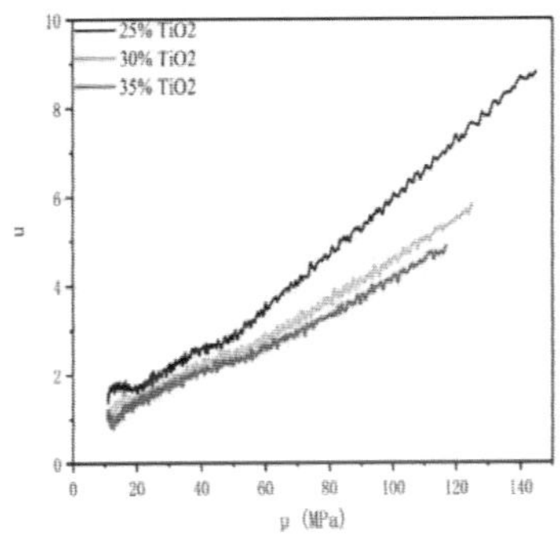

Figure 3. u-p curves

From Figure 2, it can be seen that the 35% TiO_2 of the coating liquid made the single hole propellant reaching the maximum pressure for the longest time, which increases the pressure plateau effect of the p-t curve. From Figure 3, it can be observed that under the conditions of loading density of 0.2g/L, the single hole propellant made from the coating solution with 35% TiO_2 content has the lowest combustion pressure and the lowest burning rate. Comparing with 25% TiO_2 content and 30% TiO_2 content of the coating solution. It can be concluded that when the TiO_2 content is 35%, it flame retardant effect is the best.

3.3 XRD analysis of residues after combustion of single hole propellant with different coating formulation

The 25%, 30% and 35% TiO_2 content coating liquids are made into a single-hole propellant, and doing the constant-volume combustion test. Taking the powder from this combustion for XRD test. Result diagram is shown in "Figure4", "Figure5", "Figure6", "Figure7".

IOP Conf. Series: Materials Science and Engineering **436** (2018) 012011 doi:10.1088/1757-899X/436/1/012011

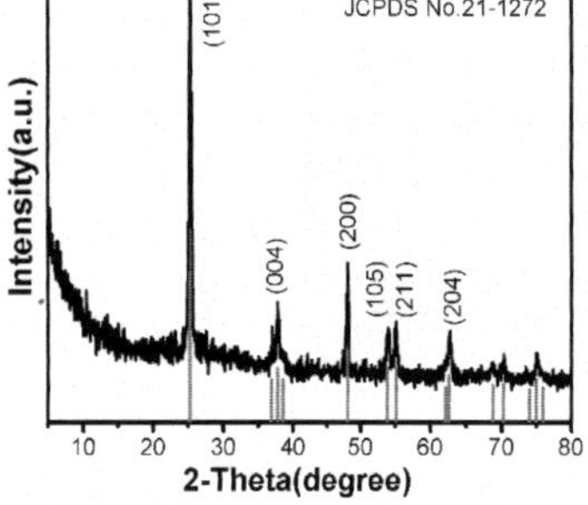

Figure 4. XRD patterns of TiO_2 for the coating solution

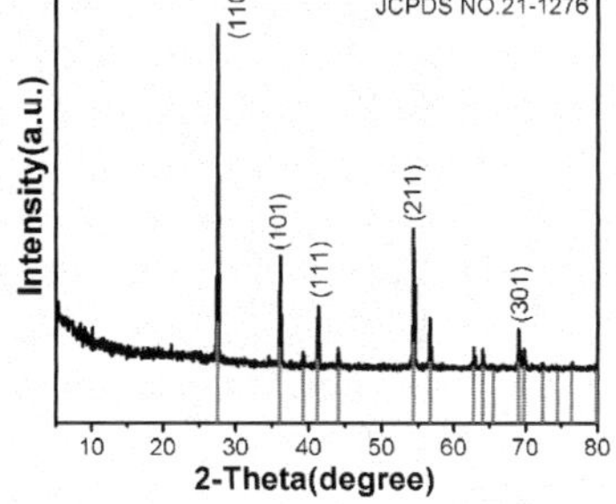

Figure 5. XRD pattern of residue after constant volume combustion test of single hole propellant containing 25% TiO_2 coating liquid formulation

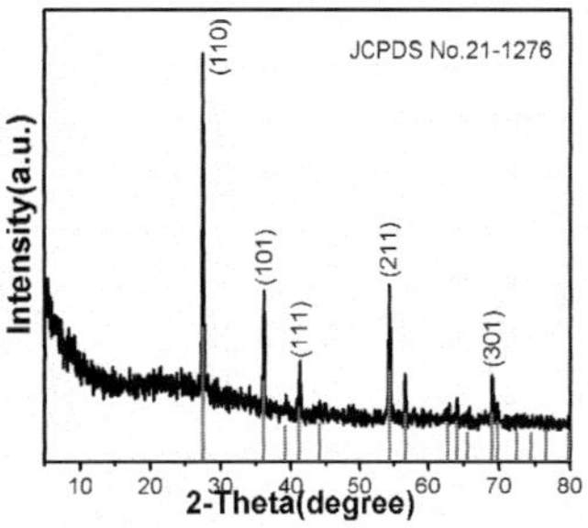

Figure 6. XRD pattern of residue after constant volume combustion test of a single hole containing 30% TiO_2 coating solution

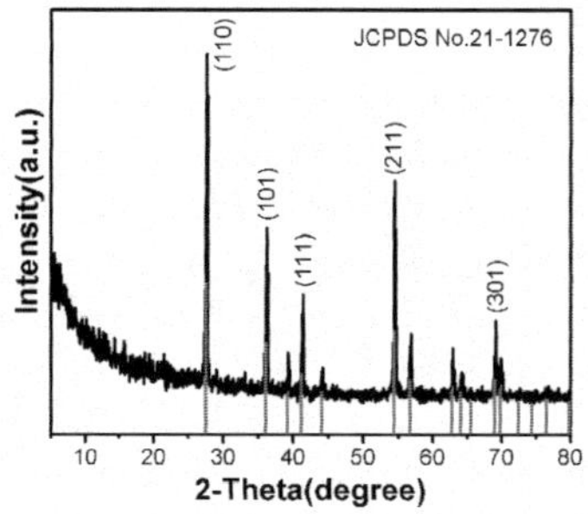

Figure 7. XRD pattern of residue after constant volume combustion test of a single hole containing 35% TiO_2 coating solution

From spectrum 4, it can be seen that the strongest diffraction peak of the sample correspond to the 101 crystal plane appearing at about 25°, which is anatase titanium dioxide. The spectrums 5, 6, and 7 are the strongest for the TiO_2 powder removed after the constant volume combustion test. The diffraction peak corresponds to the 110 crystal plane appearing at about 27°, which are rutile titanium dioxide. It can be proved that the TiO_2 crystal form has changed from anatase to rutile during constant volume combustion.

3.4 Constant volume combustion test of coating.

After sampling and burning in accordance with Table 1, p-t curves as showed in Figure 8.

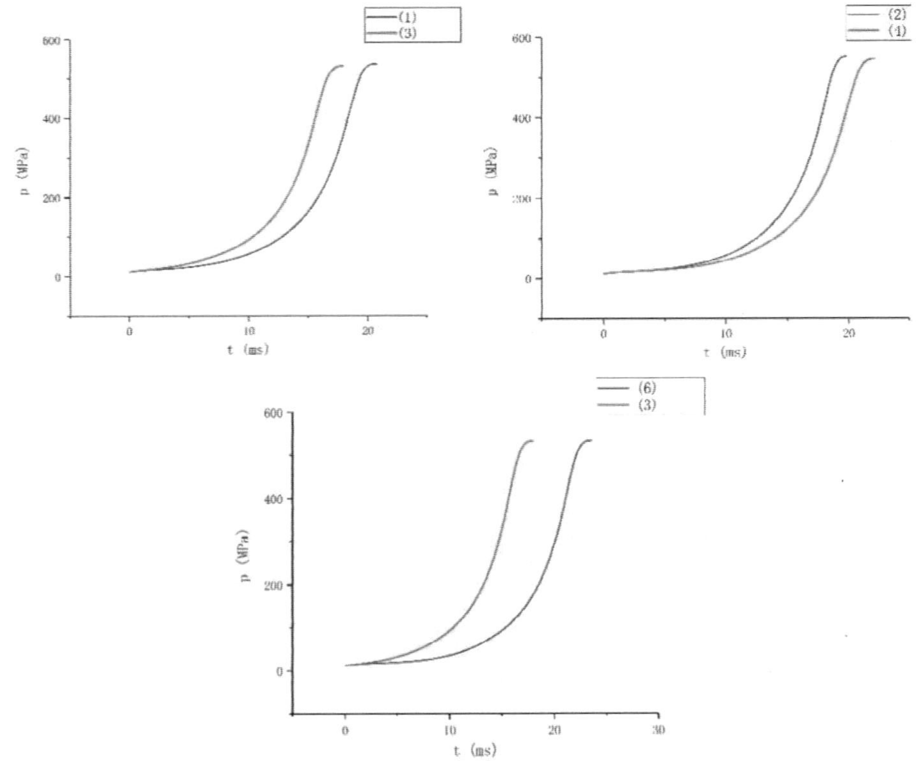

Figure 8. Constant-temperature combustion test p-t curves

Table 3. Maximum pressure and maximum pressure occurrence time of constant volume combustion test

Group number	Greatest pressure	Maximum pressure occurrence time
#1	537.36	20.756
#2	551.02	19.727
#3	532.48	17.906
#4	545.76	22.123
#5	550.06	18.215
#6	533.37	23.482

Try to ensure that each group of experimental propellants has a similar quality. The quality of the propellant will mainly affect the maximum pressure generated by the combustion. From the comparison of curves #1 and #3 in Figure 8 and Table 3, it can be seen that under similar mass conditions, the maximum pressure of the pure coated propellant is 2.85ms later than the maximum pressure time of the coated and the uncoated propellant. The increase of the maximum pressure time indicates that the coating has a role in increasing the burning of the propellant. Comparing curves #2 and #4, #3 and #6. It can be illustrated that 35% TiO_2 to 30% TiO_2 is more effective in increasing the gradual increase in the combustion of the propellant.

4. Conclusion

(1) The addition of TiO_2 significantly reduces the energy density of the single hole propellant. When

the content of TiO_2 in the coating liquid of the single hole propellant is 35%, at the same loading density, the maximum pressure reached is the minimum and the time required is the longest. The burning speed is obviously affected by the TiO_2 content. And it has a good flame retardant effect.

(2) XRD analysis of residues after combustion of single hole propellant with different coating formulation. During the process TiO_2 has undergone a crystal transformation from anatase to rutile.

(3) The coating layer plays an important role in increasing the burning of the propellant. 35% TiO_2 works better, and it is more effective in improving the gradual increase of the propellant's combustion.

References

[1] Wang Zeshan, Xu Fu-ming, Zhang Hao-xia. Design Theory of Propellant Charge[M]. Beijing: Ordnance Industry Press, 1995. 6.

[2] Du Ping. He Weidong, Wang Zeshan. Research on the flame retardants of titanium dioxide in the combustion of propellant[J]. Chinese Journal of Energetic Materials, 2005(02): 99-102-66.

[3] Luo Yun-jun, Li Feng. The propellant combustion contral technology[J]. Journal of Combustion Science and Technology, 1998,1(4): 24 -30.

[4] Du Ping. Study on the characteristics of coating and cladding interface of low-temperature propellant[D]. Nanjing University of Science and Technology, 2005.

[5] Shi Xianyang, He Weidong,Wang Zeshan. The closed vessel combustion and ballistic property study of low temperature sensitivity high energy nitroanine[J]. Journal of Ballistics, 2002, 14(1): 8-13.

[6] Wen Gang, Du Ping, Liao Xi. Determining Actual Burning Rate of Propellant with a Closed Bomb[J]. Journal of Explosives & Propellants, 2011, 34(03): 57-60.

Secondary Development and Application of Satellite Dynamic Simulation Platform Based on ABAQUS

Xiaotong Zhang[a], Wenlai Ma[b] and Xiaolei Jiao

Harbin Institute of Technology, Harbin,150001,China

[a]zhangxiaotong0627@163.com; [b]mawenlai@hit.edu.cn

Abstract. As a strong coupling system, structure of the whole satellite has too many design parameters which are undetermined before initial design. In order to overcome the existing problems such as high update frequency of related parameters, complexity of the mouse operation and the low effectiveness of manual modelling, based on secondary development of ABAQUS, a satellite dynamic simulation platform is carried out. First of all, a convenient and friendly user interface is established by writing registration file and the dialog interface creation file based on Python scripting language. Besides, kernel execution scripts is also developed based on Python scripting language, therefore, the parametric and automation modelling as well as simulating process are realized and the efficiency of analysis can be improved.

1. Introduction

In the process of launching, orbiting and returning, thus the satellite undergoes many physical and chemical changes such as high temperature, acceleration and vibration, causing excitation to the structure which directly affects the stability of the satellite structure and the accuracy of navigation and communication. Therefore, the dynamic simulation analysis of the whole satellite is the foundation to verify the correctness of the model establishment[1]. Based on the ABAQUS software, a new method of modelling and dynamic simulation is put forward in this paper, which will support the further design and optimization.

The engineers need to study dynamic characteristics of the whole satellite under the conditions of different dimensions and physical properties at the beginning of its design. However, there are too many cumbersome mouse operations when using ABAQUS for modelling, and once the model needs to be modified, the mouse operations have to be repeated. In order to avoid these unnecessary manual operations which further increase the computational cost and error probability, this paper uses the application program interface provided by ABAQUS to write a new user graphical interface and the pre and post processing kernel script in the Python language. Users only need to input the corresponding parameters in the interface, then the modelling and simulation of the whole satellite can be realized, which can effectively improve the efficiency and provide a simple, efficient and universal platform for satellite design[2].

2. Key technology of ABAQUS secondary development

This paper realizes satellite parametric modelling and simulation based on kernel script and graphical user interface. It mainly involves the following three types of files[3]: kernel execution file xx.py is the core of the plug-in, and it is used to drive ABAQUS/CAE to execute internal commands to carry out CAE modelling and data processing; registration file xx_plugin.py is used to register various kinds of control keywords, check the legitimacy of data, and register plug-in tools to the Plug-ins menu in ABAQUS/CAE; the dialog interface creation file xxDB.py is used to form a friendly input interface by defining and associating various forms and controls, so that users can manage all the parameters in a unified graphical interface. The running process of the platform is shown in figure 1.

2.1 ABAQUS/CAE kernel script

ABAQUS provides the script interface for users, and the script that is transferred will directly access the kernel script of the program, thus implementing the pre- and post-processing. This not only facilitates the management of data, but also makes it more convenient to use. Users only need to set and modify relevant parameters in the script program and execute it, and various complicated operations in the pre- and post-processing can be completed quickly and accurately. The ABAQUS kernel execution program is based on Python language[4], and its script interface involves about 500 objects which can be divided into three categories[5], as shown in figure 2. These objects in ABAQUS are crucial for the kernel script to run efficiently.

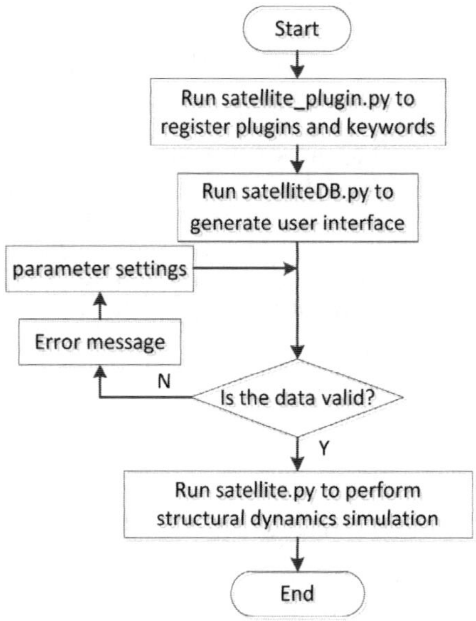

Figure 1. running process of the platform

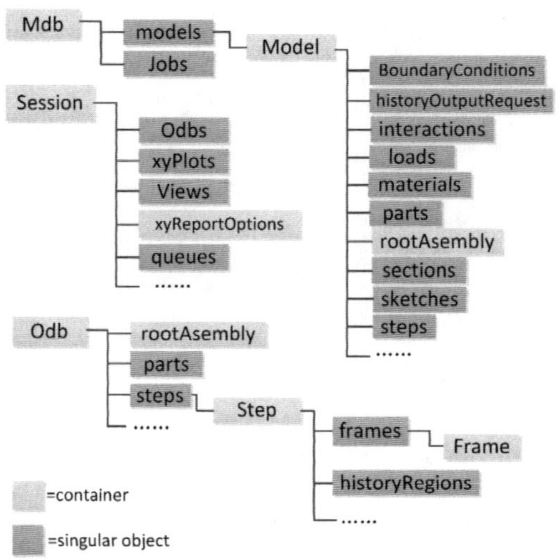

Figure 2. ABAQUS object model

2.2 ABAQUS graphical user interface

The graphical user interface mainly includes two parts: kernel process and GUI process[6]. ABAQUS realizes the human-computer interaction through the data interaction between these two parts, as shown in figure 3.There are two ways to create a GUI graphical interface based on ABAQUS/CAE: the RSG Dialog Builder and the ABAQUS GUI Toolkit. The former is embedded in ABAQUS/CAE, which provides an efficient and convenient method to create user interfaces. However it has a very limited variety of controls. In order to meet the complex needs of satellite modeling, this paper uses ABAQUS GUI toolkit to create user interfaces, which is to edit GUI commands directly in the source file. This method can realize many functions which cannot be realized by the RSG Dialog Builder[7], such as display error prompts, control usage status of the controls, etc.

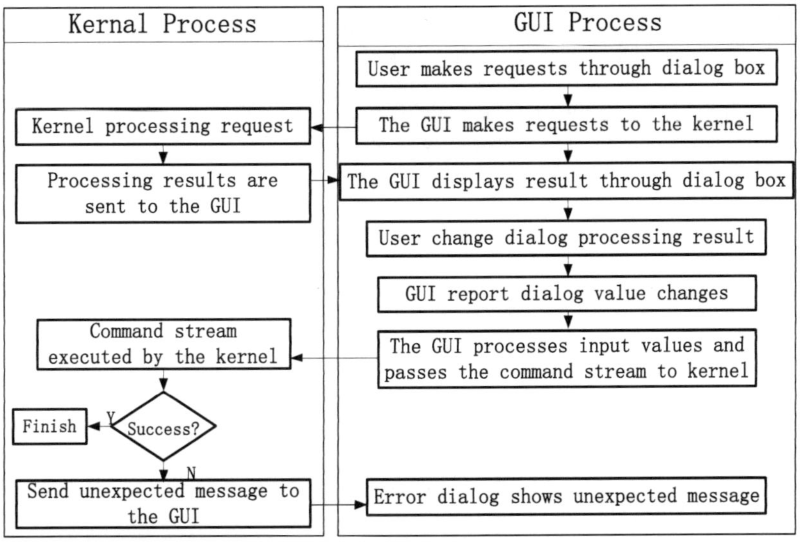

Figure 3. interaction between GUI and kernel in ABAQUS/CAE

2.3 Model simplification technology

In order to reduce weight and ensure structural safety of satellite, the honeycomb sandwich plate is applied widely in the modern spacecraft design because of its excellent capability[8]. When the finite element calculation of the satellite is carried out by the general simulation software, the lack of the honeycomb cell data library makes the processing of the honeycomb sandwich plates a challenging problem. This paper combines the secondary development of ABAQUS and the existing equivalent principle of honeycomb sandwich plates to get the equivalent model of honeycomb sandwich plates to complete the modelling and dynamic solution of the whole satellite.

At present, there are three kinds of equivalent methods commonly used in engineering: the sandwich plate theory, the honeycomb plate theory and the equivalent plate theory[9]. Here we use the equivalent plate theory as an example to describe the equivalent principle. Let's assume the total thickness of the honeycomb sandwich plate is 2H, the thickness and the density of the upper and lower plates is d and ρ_f respectively, the density of the honeycomb core is ρ_c, the elastic modulus is E, and the Poisson's ratio is μ. The honeycomb sandwich plate is equivalent to the isotropic shell element with unequal thickness. It is assumed that the equivalent plate satisfies the Kirchhoff hypothesis of the thin plate bending theory[10]. The parameters of equivalent thickness, equivalent elastic modulus, equivalent Poisson's ratio and equivalent density can be obtained as follows:

$$t_{eq} = \left[d^2 + 12\left(H - \frac{d}{2}\right)^2\right]^{1/2} \tag{1}$$

$$E_{eq} = \frac{2Ed}{t_{eq}} \tag{2}$$

$$\mu_{eq} = \mu \tag{3}$$

$$\rho_{eq} = \frac{2\rho_f d + 2\rho_c(H - d)}{t_{eq}} \tag{4}$$

3. Development of the whole satellite dynamic simulation platform

3.1 Plug-in operation

Save the prepared satellite.py, satelliteDB.py, satellite_plugin.py files and the required graphics files into the plug-in folder under the ABAQUS installation directory, and users can find the satellite parametric dynamic simulation platform plugin in the plugins drop-down menu. After that the user interface (as shown in figure 4) created by the dialog interface creation file and registration file will be displayed. In the interface, users can set the structure parameters of the satellite cabin, solar arrays, antenna and material properties according to their own needs. If the input parameters are invalid, such as the length parameter is negative, the system will pop up an error prompt dialog to remind users to input the correct and effective parameters. Click the OK button after completing the parameter input and the plug-in will automatically run the kernel program file which includes the establishment of the satellite model, the assignment of material properties, the application of constraints and loads, the automation of the grid and the corresponding simulation calculations as needed.

3.2 Geometric modelling

The plug-in kernel program performs the whole satellite modeling, material properties attributing and assembly processing after users finish inputting the parameters needed as shown in figure 4 and the material parameters shown in figure 5 input by the user. After that, users will get the whole satellite model, as shown in figure 6, some of the key codes are as follows:

p= mdb. models['Model-1'].Part(name='Part-4',dimensionality=THREE_D,type=DEFORMABLE _BODY) # create a three-dimensional model

p. Section Assignment(region=region,sectionName='Section-1',offset=0.0,offsetType=MIDDLE_ SURFACE, offset Field='',thickness Assignment=FROM_SECTION)#assign section properties

a.WirePolyLine(points=((v1[vnum1[j]],v4[vnum2[j]]),),mergeType=IMPRINT, meshable=OFF) # create a rope unit

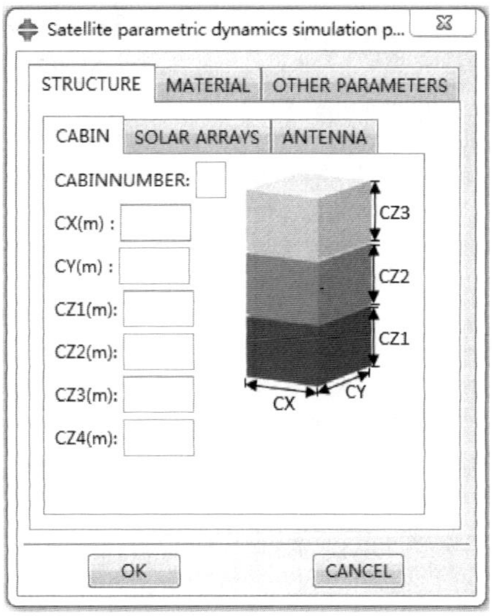

Figure 4. user interface of structure size setting

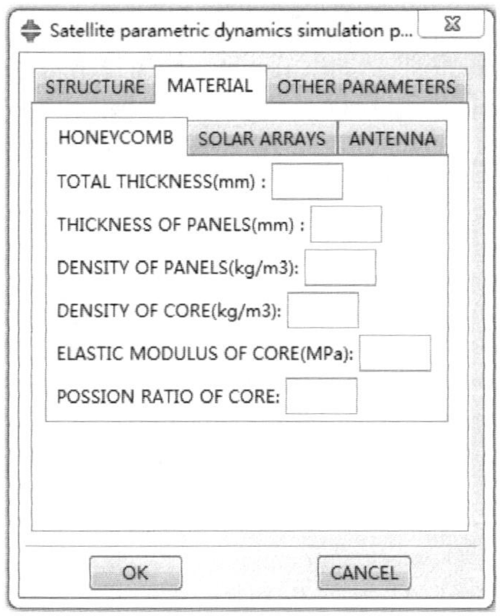

Figure 5. user interface of material property setting

3.3 Constraint exertion

After the satellite modelling is completed, the interactive constraints of different parts of the satellite are exerted, such as the rope connection between the ribs of the antenna and the connection between the cabin and the antenna. The parameters of the constraints discussed above can be set by users in the interface shown in figure 7. The model with constraints is shown in figure 8. Some of the key codes are as follows:

a.ConnectorOrientation(region=csa.getSet(),localCsys1=dl[1]) #Connection section assignment

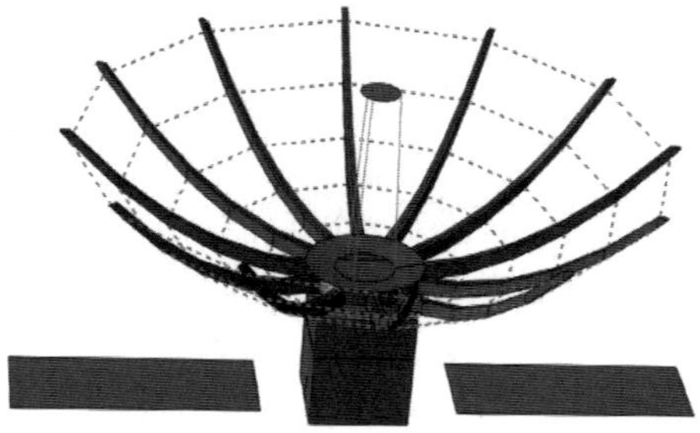

Figure 6. the whole satellite model

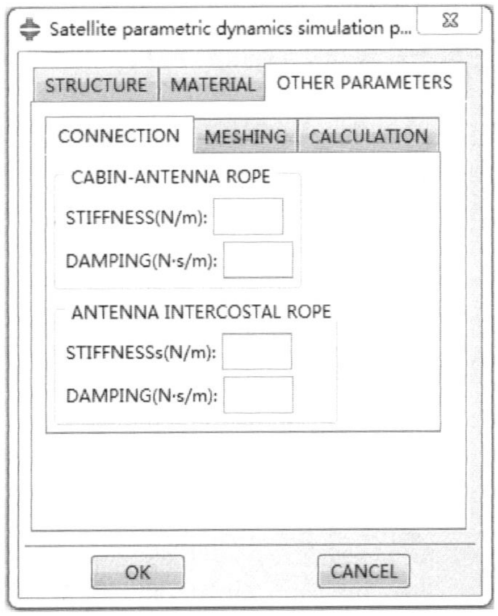

Figure 7. user interface of constraint attribute setting

3.4 Meshing

Next, the plug-in program automatically performs meshing according to the meshing parameters set by users. The finite element model of the whole satellite is shown in figure 9.

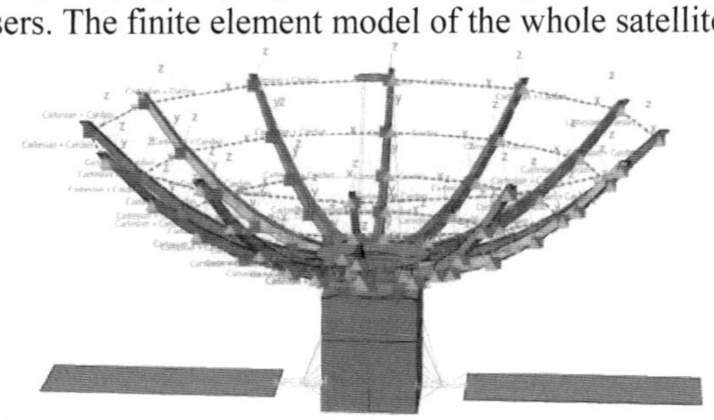

Figure 8. satellite model with constraint

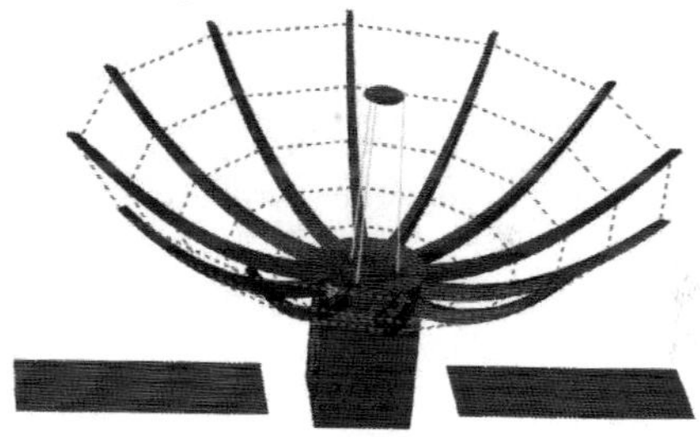

Figure 9. finite element model of satellite

3.5 Structural dynamic analysis

Modal analysis is an important approach to analyzing the dynamic characteristics of satellites, such as vibration. The modal calculation of the satellite model established in the previous section is carried out. Some of the key codes are as follows:

```
mdb.Job(name='satellite', model='Model-1',......)#create a job
mdb.jobs['satellite'].submit(consistencyChecking=OFF)#submit calculation
session.writeImageAnimation(fileName='mode-%i'%i,format=AVI,canvasObjects=(session.
viewports ['Viewport: 1'], ))# output mode shape
```

Conclusion can be drawn from the mode shapes shown in figure 10 that the equivalent displacement is mostly concentrated on the top of the antenna rib and the solar arrays, which indicates that it is necessary to pay more attention to strengthen the antenna rib and the solar arrays in the following design to improve the dynamic stability of the whole satellite.

4. Conclusion

This paper carries out the secondary development of ABAQUS based on Python language and the plug-in program is written to realize the model equivalent, model establishment and dynamic

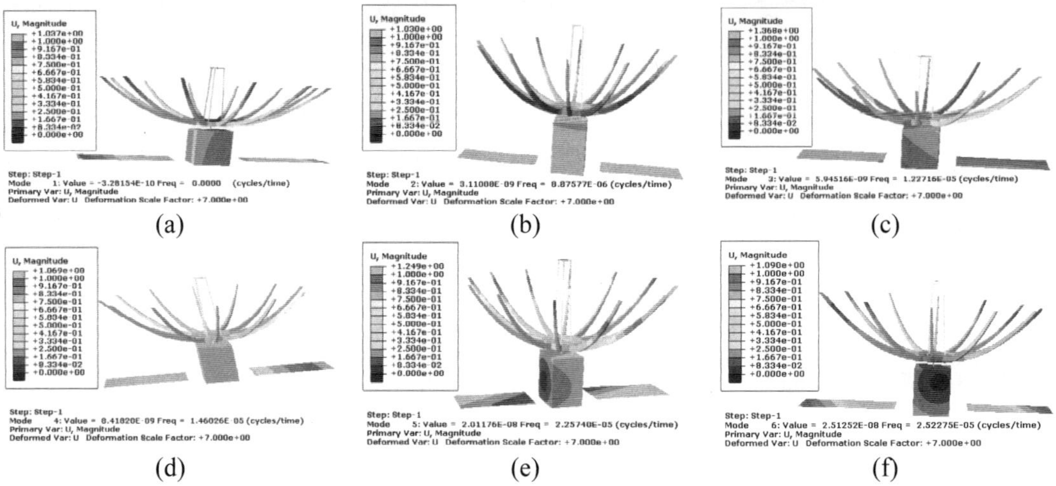

Figure 10. First 6th mode shapes

simulation of the satellite. The developed user interface is easy to operate, and enables the users who lack the experience of the ABAQUS application to carry out the dynamic simulation analysis of the whole satellite conveniently. While achieving unified management of parameters, it effectively improves simulation efficiency and avoids operation error. The plug-in can be extended to the design of the same type of satellites, and provides an effective reference for subsequent design and optimization of other types of satellites.

References

[1] Pulecchi, T., Casella, F., & Lovera, M. 2010 Object-oriented modelling for spacecraft dynamics: tools and applications. *Simulation Modelling Practice & Theory* **18(1)** 63-86.

[2] Song T, Cheng-Gong M A, Wang H, et al. 2018 Parametric modeling and harmonic response analysis of drum chest frame based on ABAQUS-Python secondary development *Wuhan. World of Building Materials.*

[3] Wang, H., and Wei, H. 2013 Application of abaqus/plug-ins in the parametric modeling and mode analysis of gun barrel. *Ordnance Industry Automation.*

[4] Luo L, Zhao M. 2011 The secondary development of ABAQUS by using python and the application of the advanced GA *Hong Kong.Physics Procedia* **22** 68-73.

[5] Zhang X D, Liu J, Zou S G, et al. 2017 Application of Abaqus secondary development to the simulation of intermediate frequency hot pipe-bending *Beijing.Journal of Plasticity Engineering.*

[6] Wang, J. L., Ping, L. I. 2009 Secondary development for gui of box girder bridge based on abaqus. *Journal of Chongqing Jiaotong University.*

[7] Meng, L. I., Cun-Gui, Y. U., Cui, E. W., and Xian-Wei, Q. I. 2013 Secondary development of abaqus and its application in creating parametric model of artillery. *Journal of Sichuan Ordnance.*

[8] Boudjemai, A., Amri, R., Mankour, A., Salem, H., Bouanane, M. H., and Boutchicha, D. 2012 Modal analysis and testing of hexagonal honeycomb plates used for satellite structural design. *Materials & Design* ***35*** 266-275.

[9] Luo, H., Liu, G., Ma, S., & Liu, W. 2011 Dynamic analysis of the spacecraft structure on orbit made up of honeycomb sandwich plates. *IEEE International Conference on Computer Science and Automation Engineering* **Vol.1** pp.83-87.

[10] Xia L J, Jin X D, Wang Y B. 2003 Equivalent analysis of honeycomb sandwich plates for satellite structure *Shanghai.Journal of Shanghai Jiaotong University* **37(7)** 999-1001.

IOP Conf. Series: Materials Science and Engineering **436** (2018) 012013 doi:10.1088/1757-899X/436/1/012013

Investigation of Notch Shear Cutting for JIS SCM420 Steel Wire Rod

P Natpukkana[1,*], S Pakinsee[1], S Boonmapat[1], P Mitsomwang[1], R Borrisutthekul[1], R Panuwannakorn[2] and L Khoa-phong[2]

[1]School of Metallurgical Engineering, Suranaree University of Technology, 111 University avenue, Muang, Nakhon Ratchasima, 30000, Thailand
[2]Thai Motor Chain Co., Ltd., 7/138 M.4, Amata City Industrial Estate, Pluakdaeng, Rayong, 21140, Thailand

*pusit@sut.ac.th

Abstract. This paper aims to study the notch shear cutting of JIS SCM420 steel wire rod. In the notch shear cutting experiment, the wire rod specimens were initially notched on the surfaces using a sharp insert installed at a lathe turning machine. Then, the specimens were cut off using the punch/die shearing tool. After the shear cutting, the sheared edge features of the wire rod were investigated and compared with the results obtained from the conventional shear cutting. In addition, the geometry of the notch, the notch depth, was varied ranging from approximately 333 to 537 μm and studied. The experimental results revealed that the cutting load resistance significantly decreased with the notch shear cutting technique. The roll over and the size of the burr zones also decreased, compared to the conventional shear cutting. This technique also improved the quality of the sheared edge in term of the roundness of the wire rod. Concerning the notch depth, it was a primary factor affecting the cutting results, especially the cutting load resistance and the occurrence of the long burr protruded from the surface of the wire rod. The improvement of the sheared edge quality by the notch shear cutting seemed to be caused by the suppression of the plastic deformation of the wire rod during shearing.

1. Introduction

Metal wire rods, especially steel wire rods, are important raw materials for large numbers manufacturing such as the production of roller chains, needle roller bearings, screws, bolts, etc. The steel wire rods are typically available in the form of long products (wire rod coils) and subjected to the punch/die shear cutting in order to cut off into designed length for their subsequent manufacturing. Normally, the edge of sheared product consists of different four zones, i.e. the roll over, the burnish, the angular fracture and the burr zones [1]. Among these, the features of the roll over and the burr zones which are resulted by the plastic deformation of the sheared rod during shearing seem to strongly affect the applications of the sheared products. For example, in the assembly of the roller chain using the steel sheared pin, when a large burr forms at the sheared edge of the pin, it cannot be assembled with the link plate of the chain.

There are researchers concentrating their studies on the features of the roll over and the burr zones with respect to the mechanical conditions of the shear cutting process. G Fang and N Hatanaga studied the effects of punch/die clearance on the sheared edge features of aluminum alloy and mild steel sheets. They showed that the roll over and the burr zones tended to be larger when increasing the punch/die clearance [2-3]. Z Tekiner studied the effects of the punch/die clearance on the cutting results of the aluminum alloy sheet. They revealed that the clearance affected the roll over, the burnish and burr zones. Namely, when the clearance increased, the roll over depth and burr length increased, while the burnish zone decreased [4]. T Kwak applied a finite element analysis to study the effects of the die clearance on the shear cutting results. From the simulation, the width and the depth of the roll over increased with the die clearance, while the burnish zone decreased with the shearing parameter [5].

In order to improve the sheared edge quality of sheet metals, especially to suppress the formation of the burr zone, the shear cutting of pre-notched sheets was carried out by some researchers. M Krinninger studied the notch shear cutting of an aluminum alloy sheet. In this study, the notches were prepared on the aluminum alloy surfaces before subjecting it to the punch/die shearing tool. From the results, the cutting load resistance and the size of the burr zone decreased when increasing the depth of the notch [6]. P Sachnik carried out the notch shear cutting of DC04 steel, 1.4301 steel, AA6014 aluminum alloy and CuSn6 alloy sheets and investigated. The experimental and simulation results revealed that the possibility of the burr formation and the burr geometries depended on the mechanical properties of the sheared sheets, the position of the notch tip and the depth of the notch [7].

From the above literature, the notch shear cutting seems to be an effective method for improving the quality of the sheared edge in term of the burr reduction. However, those research results are based on the metal sheet shearing (two-dimensional deformation) and not sufficient for understanding the features of the sheared edge of the wire rods which are deformed in three-dimension during the shear cutting. In this work, the authors aim to investigate the notch shear cutting of a wire rod. The JIS SCM420 steel wire rod was chosen as the investigation material. The features of the sheared edge were examined and compared with those obtained by the conventional shear cutting. In addition, the effects of the notch depth on the notch shear cutting results were studied.

2. Material and experimental conditions

For the shear cutting experiment, a JIS SCM420 steel wire rod which had a diameter of 4.48 mm was chosen. The mechanical properties of the JIS SCM420 steel are reported in Table 1. The shear cutting specimens were prepared to have a length of 31 mm. To investigate the notch shear cutting of the wire rod, the specimens were notched in the circumference with a sharp insert installed at a lathe turning machine.

Table 1. Mechanical properties of JIS SCM420 steel rod [8].

Properties	Metric
Young's modulus (GPa)	210
Tensile strength (MPa)	790
Yield strength (MPa)	380
Poisson's ratio	0.3

Figure 1 represents the punch/die shearing tool and the specimen configuration for the shear cutting experiment. During cutting, the punch was moved downward by the pushing head of an universal testing machine. The load cell (Capacity: ±100 kN) was installed at the machine for measuring the cutting load resistance. The main cutting tools, the punch and the die were made of a high-speed tool steel which had a hardness of approximately 62 HRC. The cutting tools were hold in the die jacket made from a plain carbon steel. The clearance between the edges of the punch and the die was adjusted to be 40 μm. The movement velocity of the punch (the cutting speed) was fixed as 5 mm·min^{-1}.

The geometry of the notch, its depth, was varied ranging from 333 to 537 μm. The variation of the notch depth was obtained by altering the feed distance of the insert into the surface of the wire rod in the notching step. Apart from the notch shear cutting, the conventional shear cutting of the wire rods was also carried out in order to examine the differences between the two cutting techniques. To do the conventional shear cutting, the wire rod specimen without notch was subjected to the same punch/die shearing tool. Other shearing parameters were the same as those explained above. The shear cutting was performed five times for each condition. After cutting, the features of the sheared edge were investigated using an optical macroscope and an image analysis software, VW-9000. In addition, to evaluate the change in the sectional profile of the wire rod due to the shear cutting, the roundness at the sheared edge of the wire rod was measured.

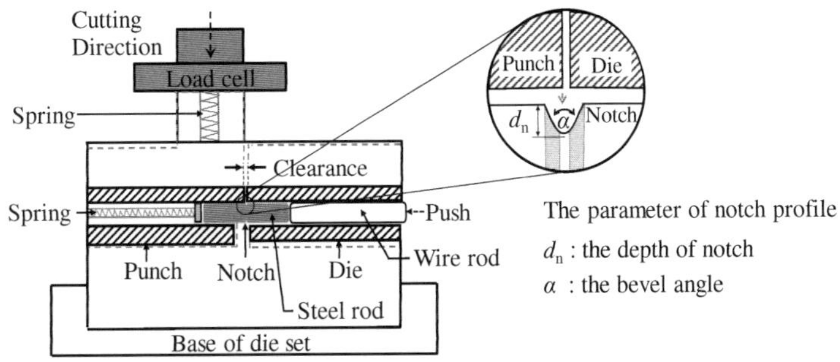

Figure 1. Experimental setup of notch shear cutting.

3. Cutting results of JIS SCM420 wire rod

3.1 Notch punch/die shear cutting versus conventional shear cutting

To investigate the notch shear cutting of the steel wire rod, the notch, which had a depth d_n and a bevel angle α of 426 μm and 50° was made on the surface of the wire rod before shearing.

3.1.1 Cutting load resistance of wire rod Figure 2 shows the relationship between the cutting load resistance (f) and the indentation depth of the punch (d) for the notch and the conventional shear cutting. Here, the indentation depth d was defined to be zero, when the surface of the punch touched the upper surface of the specimen during the shear cutting. As seen in this figure, the cutting load resistance at the shallow indentation depth, $0 < d < 0.4$, was similar for the two cutting cases. Seeing the peak load resistance, the maximum load resistance (f_{Max}) in the case of the notch shear cutting was significantly lower than that of the conventional shear cutting case. However, the peak position (d_{Peak}) of the cutting load resistance was similar for both cutting cases. Considering the position where the wire rod separated ($d_{Sep.}$), it was seen that the separation position was earlier when applying the notch shear cutting technique. In addition, from the characteristic of the cutting load, the drop of the load resistance from the peak point for the two shearing cases was fairly similar. This seemed to indicate that the notch on the wire rod surface did not completely change the fracture behavior of the sheared wire rod. However, the notch resulted in the reduction of separation resistance, remarkably.

3.1.2 Sheared edge features of wire rod Figure 3 shows the side-view images of the sheared rods from the conventional and the notch shear cutting. As seen in this figure, the large roll over zone was observed in the case of the conventional shear cutting, while this zone did not form in the case of the notch shear cutting. Typically, the roll over is caused by the plastic deformation of the rod before its separation. Since the notch made on the wire rod surface reduced the separation resistance of the rod, the plastic deformation during shearing reduced. The reduction of the plastic deformation seemed to remarkably suppress the formation of the roll over.

IOP Conf. Series: Materials Science and Engineering **436** (2018) 012013 doi:10.1088/1757-899X/436/1/012013

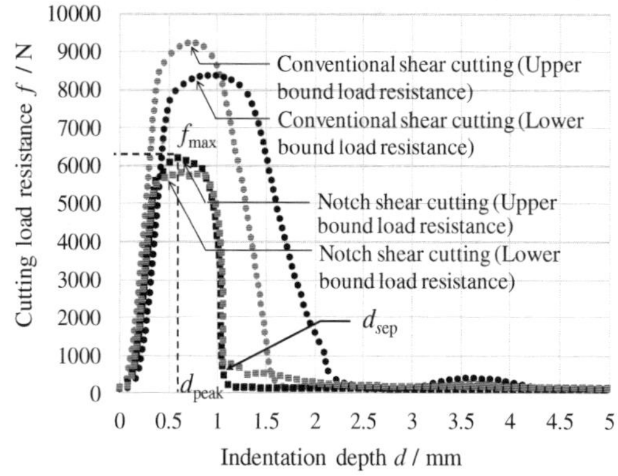

Figure 2. Load resistance of steel rod for conventional and notch shear cutting.

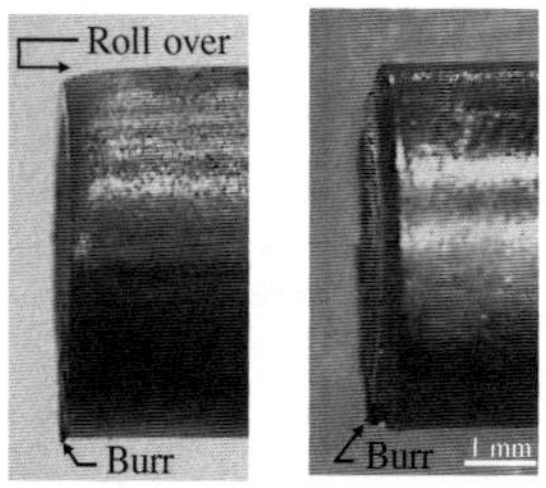

(a) Conventional shear cutting (b) Notch shear cutting

Figure 3. Side-view images of sheared wire rods cut by conventional and notch shear cutting.

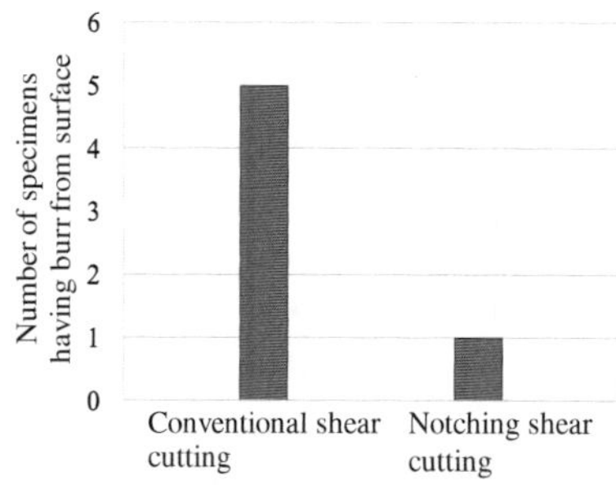

Figure 4. Number of specimens having burr protruded from surface of wire rod for conventional and notch shear cutting.

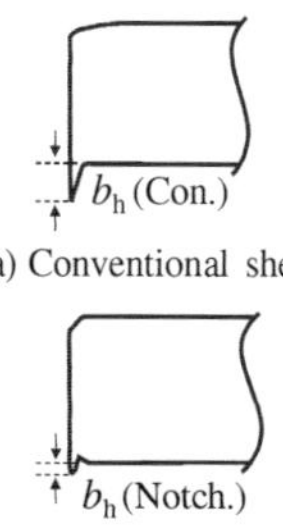

(a) Conventional shear cutting

(b) Notch shear cutting

Figure 5. Definition of $b_{h(Con.)}$ and $b_{h(Notch.)}$.

Apart from the difference of the roll over zone, seeing the burr at the sheared edge, it was found that, in the case of the conventional shear cutting, the long burr protruded from the surface of the wire rod formed, as seen in figure 3 (a). In the case of the notch shear cutting, as seen in figure 3 (b), the burr at the sheared edge seemed to still generate, but it was shorter than the depth of the notch and did not protrude from the surface of the wire rod.

Next, the authors furthermore investigated the occurrence of the burr protruded from the surface of the wire rod. Figure 4 shows the number of the specimens having the burr protruded from the surface of the wire rod for the conventional and the notch shear cutting. As shown in this figure, the protruded burr was observed in all the wire rods sheared with the conventional shear cutting, while it was observed at only one wire rod cut using the notch shear cutting technique.

The geometry of the burr was furthermore investigated. Here, the length of the burr for both shear cutting cases ($b_{h(Con.)}$ and $b_{h(Notch.)}$, as defined in figure 5) were measured. In the case of the conventional shear cutting, the length of the burr protruded from the surface of the wire rod $b_{h(Con.)}$ was measured as 89.73 μm (160.05 ~ 63.22 μm), while the length of the protruded burr $b_{h(Notch.)}$ for the case of the notch shear cutting was significantly shorter. Namely, it was approximately 26.70 μm.

From figures 3 and 4, the notch shear cutting technique was confirmed to be an effective technique for improving the quality of the sheared edge, especially in terms of the reductions of the roll over size, the possibility of the long burr formation, and the length of the burr protruded from the surface of the sheared wire rod.

3.1.3 Roundness of sheared wire rod The roundness at the sheared zone is an important property of the sheared wire rod. It is always considered, when the sheared rods are assembled with small holes or other components. Therefore, in this study, the authors investigated the roundness of the sheared wire rod when applying the notch shear cutting technique. To evaluate the roundness of the sheared edge, the images of the sheared surfaces from both shearing cases, as shown in figures 6 (a) and (b), were prepared. Then, the vertical and the horizontal distances of the sheared edge, D_y and D_x shown in figure 6 (c), were measured.After that, the D_y/D_x ratio was calculated. In the case of the sheared specimens having the burr protruded from the surface of the wire rod, the length of the protruded burr was excluded from the distance D_Y before calculating the D_y/D_x ratio.

Figure 7 shows the roundness at the sheared zone of the wire rod, compared to the roundness of the steel wire rod before shear cutting. From this figure, it was found that, in the case of the conventional shear cutting, the average D_y/D_x ratio was 0.977. This value indicated that the wire rod tended to be laterally deformed during shearing. In this shearing case, the lateral deformation deteriorated the roundness of the sheared wire rod. Considering the case of the notch shear cutting, the average D_y/D_x ratio was nearly 1.00 and equal to that of the raw wire rod. It was indicated that the wire rod almost did not deform in the both directions. Then, the roundness of the sheared wire rod could be kept to be the same as that of the raw material.

For the case of the conventional shear cutting, since the separation resistance of the wire rod was higher, the sheared zone of the wire rod underwent the large amount of the plastic deformation before the separation. This seemed to be a mainly cause of the poor roundness. On the other hand, in the notch shear cutting case, the notch made on the steel wire rod reduced the separation resistance.Then, the plastic deformation at the sheared zone during the shearing reduced. As a result, the roundness of the sheared zone was kept and superior, compared to that of the conventional shear cutting case.

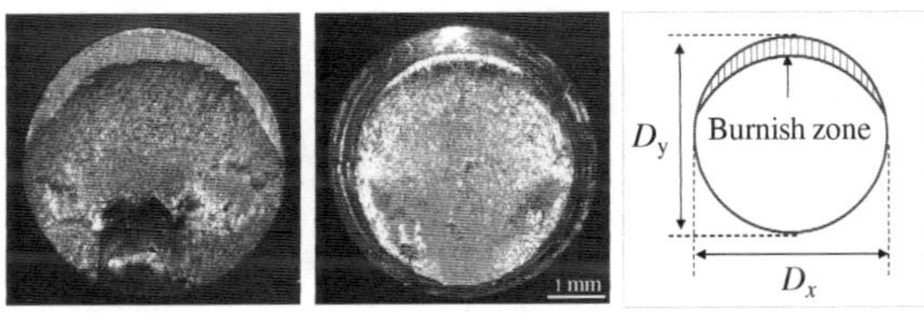

(a) Conventional shear cutting (b) Notch shear cutting (c) Measurement distance

Figure 6. Sheared surface of wire rod and measurement distances for roundness.

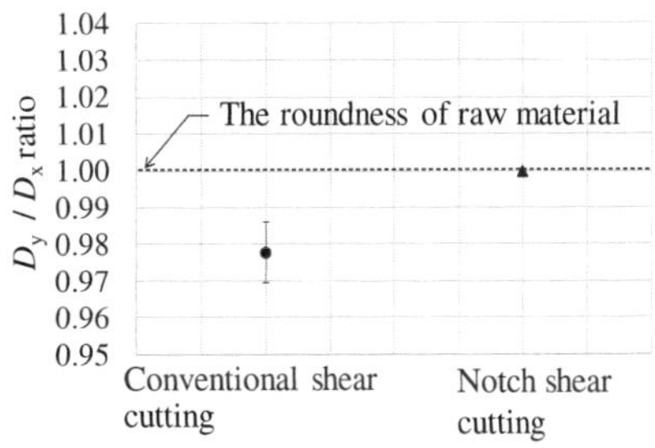

Figure 7. Roundness at sheared edge of wire rod for conventional and notch shear

3.2 Effects of notch depth on notch shear cutting results of wire rod

After knowing the effectiveness of the notch shear cutting for the wire rod, the effects of the notch geometry, the depth of the notch was varied and investigated. To vary the depth, the feed distance of the insert into the wire rod was altered as 750, 950 and 1000 μm. However, since the wire rod elastically deformed during making the notch, the actual depth of the notch was slightly lower than the feed distance. Namely, the actual depths d_n of 333, 426 and 537 μm were obtained from the applied feed distances. For this investigation, the bevel angle of the notch was fixed to be 40°.

3.2.1 Effects of notch depth on cutting load resistance of wire rod Figure 8 shows the cutting load resistance when varying the notch depth of 333, 426 and 537 μm. It was seen that the f_{Max} was significantly decreased with increasing the depth of the notch. For the position d_{Peak}, it occurred at a shallower indentation depth when increasing the notch depth. Seeing the separation position of the wire rod, it appeared to be invariant when varying the depth of the notch. From this cutting load resistance, it was indicated that the separation resistance of the wire rod significantly decreased with increasing the notch depth. However, from the characteristic of the load drop, the final fracture of the wire rod did not change with the notch depth.

IOP Conf. Series: Materials Science and Engineering **436** (2018) 012013 doi:10.1088/1757-899X/436/1/012013

3.2.2. Effects of notch depth on sheared edge features of wire rod Figure 9 shows the representative side-view photographs of the sheared edges for the cases of d_n = 333, 426 and 537 μm. As seen in this figure, the roll over was not observed for all notch depths. Next, seeing the lower side of the sheared edge, for the shallow notch, d_n = 333 and 426 μm, the burr protruded from the wire rod surface seemed to be observed. However, when the notch depths increased to 537 μm, the protruded burr tended to be absent.

Figure 10 represents the possibility of the protruded burr occurrence when varying the notch depth. From this figure, it was found that the possibility of the protruded burr formation decreased remarkably, when the notch depth increased from 333 μm to 426 and 537 μm. From the protruded burr length measurement, the tendency of the length with respect to the notch depth was not observed. Here, the average length $b_{h(Notch.)}$ was 36.77 μm (50.73 ~ 22.48 μm), 48.46 μm (63.45 ~ 33.48 μm) and 34.33 μm (40.43 ~ 28.23 μm) for the notch depths of 333, 426 and 537 μm, respectively. From the non-monotonic tendency of the protruded burr length, it indicated that there was a large variation of the protruded burr length, when varying the notch depth.

3.2.3. Effects of notch depth on the roundness of wire rod Figure 11 shows the sheared surface of the wire rod for the difference notch depth, while figure 12 represents the plot of the D_y/D_x ratio for various notch depths. From figure 12, the ratio was almost equal to one for all notch depth conditions. It was indicated that the roundness of wire rod could be kept and was not deteriorated due to the shear cutting, when varying the notch depth.

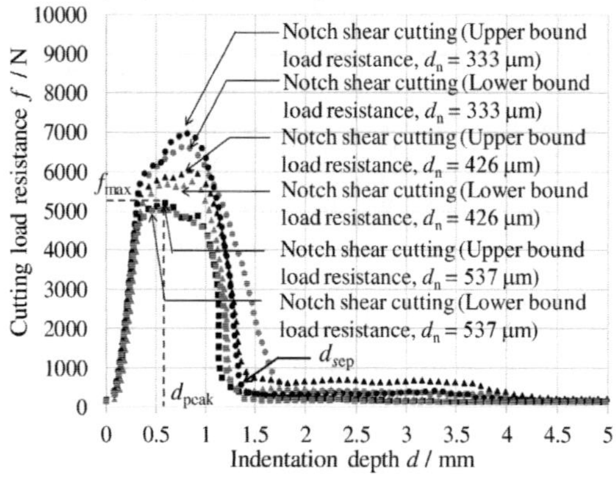

Figure 8. Load resistance of steel rod cut by various depths of notch ($\alpha \approx 40°$).

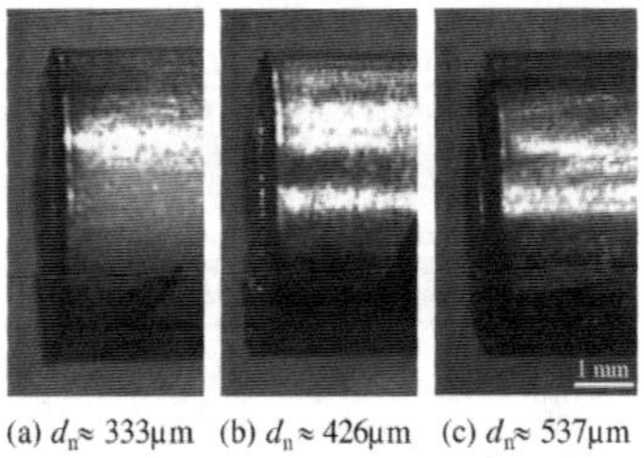

Figure 9. The Side-view photographs of sheared wire rods for depth of notch ($\alpha \approx 40$ °).

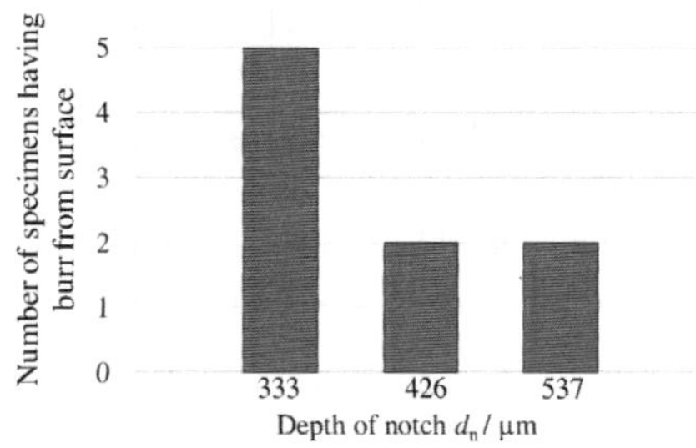

Figure 10. Number of specimens having burr protruded from surface of wire rod for notch depth.

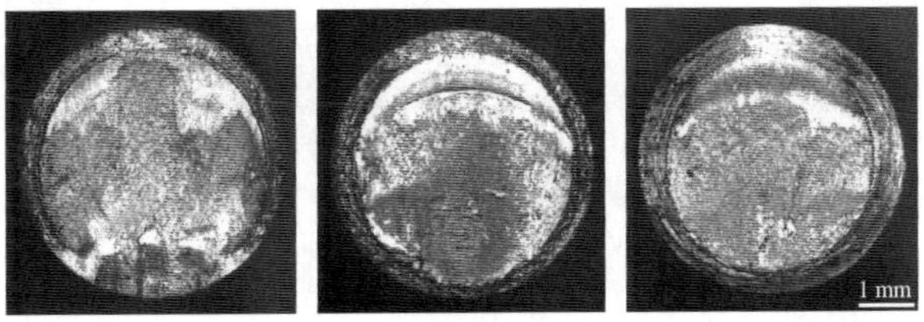

(a) $d_n \approx 333\mu m$ (b) $d_n \approx 426\mu m$ (c) $d_n \approx 537\mu m$

Figure 11. Sheared surface of wire rod for notch depth.

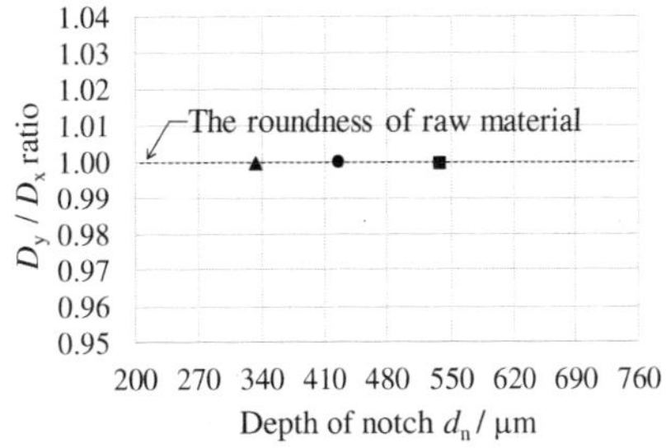

Figure 12. Roundness at sheared edge of wire rod for notch depth.

4. Conclusions

1) The notch shear cutting remarkably improved the quality of the sheared wire rod in terms of the roll over, the protruded burr length reductions and the keep of the roundness of the wire rod. Also, the possibility of the protruded burr formation decreased by the technique.
2) The sheared edge improvement of the notch shear cutting seemed to be resulted by the small plastic deformation of the wire rod during shearing. The plastic deformation remarkably suppressed due to the low separation resistance of the pre-notched wire rod.
3) The roll over at the sheared edge of the wire rod did not occur for all notch depths.
4) When the notch depth was varied, the possibility of the protruded burr formation tended to decrease. However, the tendency of the burr length with respect to the notch depth was not observed.
5) The roundness of the sheared wire rod could be kept and was similar when varying the notch depth.

References

[1] Natpukkana P, Mitsomwang P, Borrisutthekul R, Panuwannakorn R and Khoa-phong L 2018 *Proc. Int. Conf.(Bangkok)* pp 20–26
[2] Fang G, Zeng P and Lou L 2002 *J. Mat. Pro.Tech.* **122** pp 249–54
[3] Hatanaka N, Yamaguchi K and Takakura N 2003 *J. Mat. Pro.Tech.* **139** pp 64–67
[4] Tekiner Z, Nalbant M and Gürün H 2006 *Mat. Design.* **27** pp 1134–38
[5] Kwak T.S., Kim Y.J. and Bae W.B. 2002 *J. Mat. Pro.Tech.* pp 462–68
[6] Krinningera M, Feistlea M, Gollea R and Volka W 2017 *Procedia Eng.* **183** pp 53–58
[7] Sachnik P, Hoque S and Volk W 2017 *J. Mat. Pro.Tech.* **249** pp 229–45
[8] Lee K.O., Kima J.M., Chin M.H.and Kang S.S. 2007 *J. Mat. Pro.Tech.* **182** pp 65–72

Effects of Temperature Variation on Mechanics Field at Crack Tip of Stress Corrosion Cracking

Mo Jinming

Chongqing College of Humanities, Science and Technology, Hechuan, Chongqing 401524, China

E-mail:1028313467@qq.com

Abstract. Creep strain near crack front zone is one of the main driving forces in nuclear structural materials of stress corrosion crack propagation. To understand the effect of temperature variation on mechanical field near crack tip of nuclear structure material in high temperature water environment, compact tensile specimen was taken as an research object and ABAQUS finite element analysis software was also adopted, the Mise stress, plastic strain and creep strain near SCC crack tip region was studied. Results show that the area of the high-stress region decreases gradually with the increasing of the creep temperature, and its declining rate was faster and faster. The effect of temperature variation on the plastic strain is not very obvious. Temperature only has an obvious effect on a small scale near crack tip. With the increase of the distance from the crack tip, the effect of temperature on creep is gradually reduced.

1. Introduction

Stress corrosion cracking (SCC) is a failure mechanism that is caused by environment, susceptible material and tensile strain of nuclear power structural materials under high temperature and high pressure aqueous environments [1-3]. During the process of stress corrosion cracking, creep occurs at high stress area near crack tip of the structural material in the high temperature water environment, which can also induced the rupture of oxide film[4], so it is becoming very important to study the creep phenomenon of nuclear welding materials under high stress conditions. MAO Xueping conducted a high-temperature creep test on the nickel-based alloy C276 as a fuel cladding material, and creep process were studied under different temperatures and stresses [5]. An Junchao studied the creep crack propagation characteristics of P92 steel, and found that the creep crack growth rate in HAZ of welded joint is larger than the creep crack growth rate of base metal [6]. Han Ningning studied the creep behaviour of A508 steel at 450-600℃, and creep constitutive equation was obtained[7].

Since stress and strain field near crack tip can greatly influence the crack growth, it is necessary to study the influence of temperature variation on crack tip creep stress and strain field in detail. The finite element model was established in this paper, and stress and strain field are analyzed. Research result provides a theoretical basis for predicting the stress corrosion cracking rate of nuclear structural materials in high temperature water environment [8].

2. Finite element modeling

2.1. Specimen model and mesh model

The standard compact tensile test specimens (1T-CT) was adopted in ABAQUS in the paper, the specimen geometry and experimental process was conducted in accordance with ASTM399 standard [9], which was shown in Figure 1, the value of specimen width W=50 mm, value of the crack length a = 0.5W, and the value of B/W is 0.25. To eliminate the stress singularity caused by the crack tip [10], the blunt circle radius at crack tip is taken as 0.5μm in numerical simulation. Due to the symmetry of the specimen, half model is adopted, as shown in Figure 2. To improve the analytical precision at the crack tip, the sub-model technique was employed to calculate distribution of the creep mechanical field at crack tip. The 8-node biquadratic plane strain element is adopted in ABAQUS finite element analysis, with a total number of the global number is 24525, as shown in Figure 2. The element type of the sub-model is consistent with whole model, and the minimum size of the element is about 0.5μm, which is shown in Figure 2.

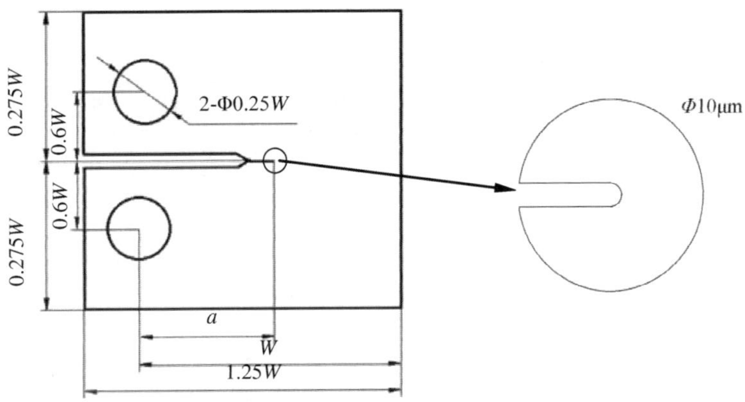

Figure 1. The geometry of the compact tensile specimen and the schematic view of the crack tip

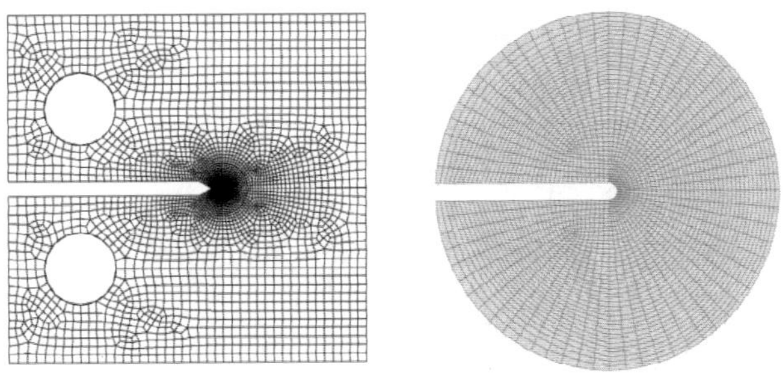

(a) Global mesh model (b) Mesh at crack tip zone

Figure.2 Finite element mesh model of 1T-CT

2.2 Material model

The austenitic stainless steel A508 is a power-hardened material whose constitutive relationship is usually characterized by Ramberg-Osgood relationship [11], and its Common calculation formula form is as follows:

$$\frac{\varepsilon}{\varepsilon_0} = \frac{\sigma}{\sigma_0} + \alpha\left(\frac{\sigma}{\sigma_0}\right)^n \tag{1}$$

Where ε is the strain, including elastic and plastic strain. σ is the total stress; ε_0 is the yield strain of the material, σ_0 is the yield stress of the material and its value is 522MPa, and n is the strain hardening exponent of the material and its value is 7.04, α is the offset coefficient of the material and its value is 1.

The creep rate is calculated using the power law model [7] in the FEM calculation:

$$\dot{\varepsilon}_{cr} = A\sigma^n t^m \tag{2}$$

where $\dot{\varepsilon}_{cr}$ is the creep strain rate. σ is the applied stress; A is the power law multiplier.

Where $\dot{\varepsilon}_{cr}$ is the creep strain rate, σ is the applied stress, A is the power law multiplier.

To understand the effect of creep temperature variation on the stress and creep strain near crack tip, The value of temperature T is 450°C, 500°C, 550°C, respectively. The creep mechanical properties of the material under different temperature is given in Table 1.

Table 1. Creep mechanical property of A508 under different temperature

Material	T/°C	A	n	m
A508	450	2.81e-6	1.47771	0.29161
	500	2.84e-5	1.17955	0.32799
	550	7.68e-5	1.2052	0.42591

2.3. Loading and boundary conditions

Loading is applied to the load hole by using the multiple point constraints facility in ABAQUS, which joins the hole centre to the nodes of the hole surface. The concentrated force is applied on the reference point so as to ensure the crack tip stress intensity factor K_I=8 MPa·m$^{1/2}$.

Two analysis steps were set up in the analysis process, the first is the process of loading on the specimen, making the sample occur elastic plastic deformation; the second stage is keeping the load constant, set the creep equivalent time 5000h.

3. Results and Discussion

3.1. Effect of creep temperature on Mises stress at SCC crack tip zone

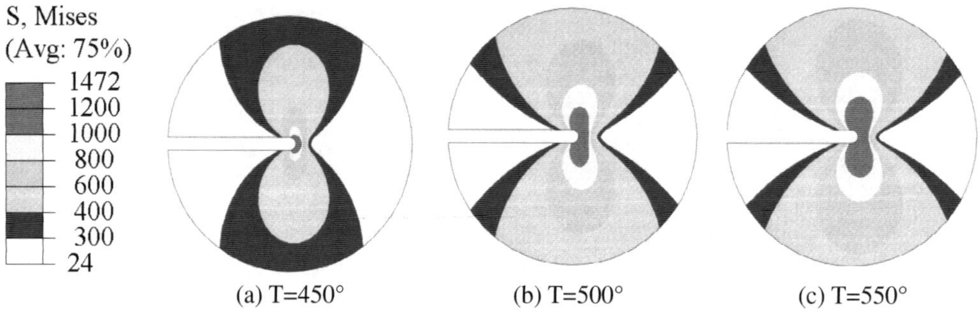

Figure 3. Comparison of Mises stress distribution after creep at different temperatures (MPa)

Mises stress distribution around crack front under different temperatures near crack tip region is shown in Figure 3. Figure 3 shows that the area of the high-stress region increases gradually with the increasing of the creep temperature, this is mainly due to the effect of creep. It can be seen that the different creep temperatures have a great influence on the Mises stress distribution around the crack tip. The higher the creep temperature is, the greater the stress decrease gradient in the crack tip area, and the creep phenomenon becomes more obvious at this moment.

3.2 Effect of Creep Temperature on PEEQ at SCC crack tip zone

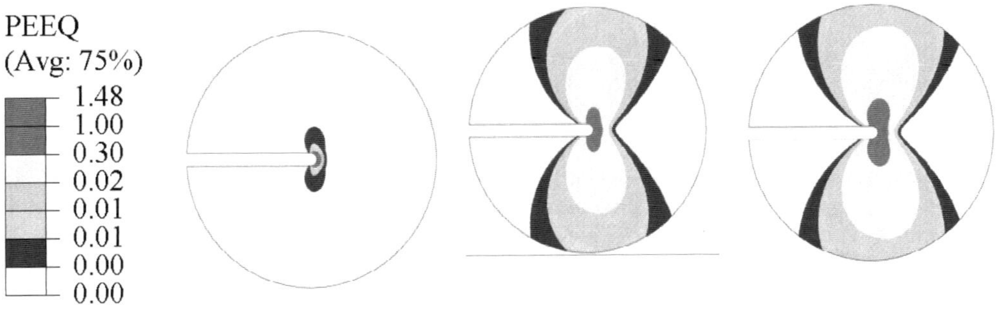

(a) T=450° (b) T=500° (c) T=550°

Figure 4. Comparison of PEEQ distribution after creep at different temperatures

Figure 4 shows the variation of equivalent equivalent plastic strain PEEQ near crack tip region, which shows that the value of equivalent equivalent plastic strain PEEQ near the crack tip region also increases with the increasing temperature under the same equivalent plastic strain scale range. It can be concluded that the change of temperature change the distribution of crack tip equivalent PEEQ after creep, and temperature also has great effect on the crack tip equivalent plastic strain.

3.3 Effect of Creep Temperature on CEEQ at SCC crack tip zone

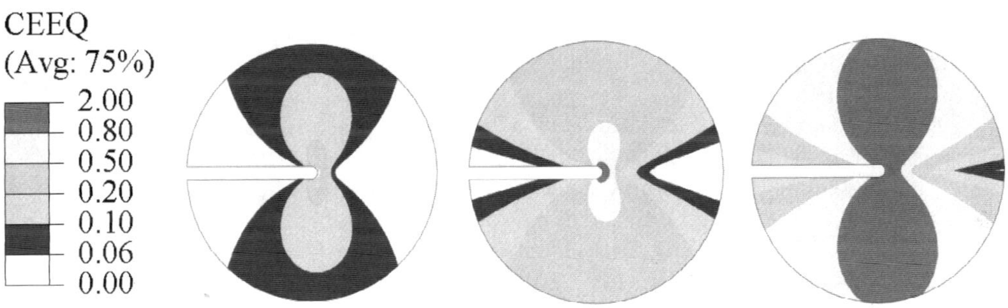

(a) T=450° (b) T=500° (c) T=550°

Figure 5. Comparison of CEEQ distribution after creep at different temperatures

Figure 5 shows the change of the equivalent equivalent creep strain CEEQ near the crack tip region under different temperatures. It can be seen that the value of equivalent creep strain CEEQ decreases with the increasing distance from the crack tip. Comparison with the cloud distribution of different temperature, equivalent creep strain is mainly concentrated in the crack tip region, and the value of CEEQ decreases with increasing distance ahead of the crack tip. When the distance from the crack tip increases, the effect of temperature on creep gradually weakens, and the variation of equivalent equivalent plastic strain CEEQ is not obvious when the distance is larger ahead from crack tip.

4. Conclusions

(1) The area of high stress zone increases with the increasing of creep temperature, and the rate of descent is faster and faster than previous, that is, the higher the temperature is, the more obvious the creep phenomenon is, Sharp angle change and change.
(2) Temperature has also great effect on the PEEQ distribution of the equivalent plastic strain at the crack tip. The value of PEEQ near the crack tip region also increases with the increasing temperature.

(3) Equivalent creep strain is mainly concentrated in the crack tip region, and the value of CEEQ decreases with increasing distance ahead of the crack tip.

References

[1] Ma Cheng, Peng Qunjia, Han Enhou. Review of stress corrosion cracking of structural materials in nuclear power plants [J]. Journal of Chinese Society for Corrosion and Protection, 2014, 34(1): 37-45.

[2] O. K. Chopa, H. M. Chung, and etc. Current research on environmental assisted cracking in light water reactor environments, Nuclear Engineering and Design, 194, 205-223, 1999.

[3] Andresen P L, Karen G, Lawrence N J. Stress Corrosion Cracking of Sensitized Type 304 Stainless Steel in 288C Water: A Five Laboratory Round Robin[M]. John Wiley & Sons, Inc. 2013. Andresen P L, Karen G, Lawrence N J. Stress Corrosion Cracking of Sensitized Type 304 Stainless Steel in 288C Water: A Five Laboratory Round Robin[M]. John Wiley & Sons, Inc. 2013.

[4] Huang X, Zhang Y, Mei M, et al. A quantitative prediction model of SCC rate for nuclear structure materials in high temperature water based on crack tip creep strain rate[J]. Nuclear Engineering & Design, 2014, 278(8):686-692.

[5] Mao Xueping, Guo Qi, Hu Suyang. Creep behaviours of ni-based alloy C276 at high temperature. Proceedings of the CSEE, 2012, 32(11):100-105.

[6] An Junchao. Research on Creep Crack Propagation Rate in HAZ of P92 Steel Welded Joint[D]. Tianjin University, 2008.

[7] Han Ningning. Study on creep behaviour of metallic materials used in oil pipeline and nuclear power[D]. Shanghai Jiao Tong University, 2015.

[8] Xue He, Xue Xiaofeng, Tang Wei. Analysis on mechanical property of oxide film in stress corrosion cracking tip of nickel-based alloys[J]. Rare metal materials and engineering, 2011, 40(7): 1188-1191.

[9] ASTM Standard E399. Standard Test Method for Linear-Elastic Plane-Strain Fracture Toughness K1C of Metallic Materials[J]. 2012.

[10] Dassault Systèmes Simulia Corp. Abaqus analysis user's manual 6.12[EB/CD], 2012.

[11] EDA Y, SHI Y, SUN S, et al. Effect of crack depth and strength mis-matching on the relation between J-integral and CTOD for welded tensile specimens(mechanics, strength & structure design) [J]. Transaction of JWRI, 1997, 26(1): 133-140.

Study of the Electrochemical Performance of Activated Carbon Bulky Paper Electrode for Electrical Double Layer Capacitor (EDLC)

Luhua Cheng[1,2], Xiaosong Du[1,*], Yadong Jiang[1] and Alexandru Vlad[2]

[1]State Key Laboratory of Electronic Thin Films and Integrated Devices, University of Electronic Science and Technology China(UESTC), Chengdu 610054, China.
[2]Institute of Condensed Matter and Nanosciences, Université Catholique de Louvain, 1348 Louvain-la-Neuve, Belgium.

*xsdu@uestc.edu.cn

Abstract. Supercapacitor have attracted much attention due to its fast power supply, long cycle life, simple principle, etc. Supercapacitors are widely used in consumer electronics. Activated carbon is a typical material for electric double layer capacitor. Here, we study the electrochemical performance of an activated carbon (TF-B520) bulky paper electrode made by vacuum filtration method. Cyclic Voltammetry and Constant Current Charge/Discharge were used to test the electrode capacitance and the rate performance. The results showing that TF-B520 activated carbon balky paper electrodes have the highest capacitance of 120F/g at scan rate 1mV/s and good capacitor behaviour at scan rate below 20mV/s and current density below 250mA/g.

1. Introduction

Reducing fossil fuel consumption and the effects of greenhouse gas emissions into the atmosphere has become global objectives being recognized as an imperative for the sustainable development of economy and society. [1-3] There is an urgent need for efficient, clean, and sustainable sources of energy, as well as new technologies associated with energy conversion and storage. Electricity production from renewable energy sources and further improvement of energy efficiency are the most promising solutions to reach these objectives. [4,5] Supercapacitors, also known as electrochemical capacitors or ultracapacitors, have attracted much attention because of their fast power supply, long cycle life, simple principle, and high dynamic of charge propagation. Currently, supercapacitors are widely used in consumer electronics, memory back-up systems and industrial power and energy management. on the basis of the energy storage mechanism, supercapacitors can be classified into electrical double layer capacitor (EDLC) and pusdocapacitor. [6-8] For electrical double layer capacitor (EDLC), the capacitance comes from the pure electrostatic charge accumulated at the electrode/electrolyte interface, therefore it is strongly dependent on the surface area of the electrode materials that is accessible to the electrolyte ions. [9] Activated carbon is one of the most widely used material for electrical double layer capacitor, because of its high surface area and low cost. In this article, we study the electrochemical performance of a kind of activated carbon form MTI, TF-B520, the activated carbon was first mixed with Multi-walled carbon nanotubes (MWCNTs) to increase the conductivity, and then made into a free standing bulky paper by vacuum filtration method, cut into

IOP Conf. Series: Materials Science and Engineering **436** (2018) 012015 doi:10.1088/1757-899X/436/1/012015

free standing electrodes, and finally assembled into a symmetrical electrical double layer coin cell for electrochemical measurements.

2. Experimental

TF-B520 activated carbon (from MTI) and MWCNTs (MTI) were used to assemble bulky paper electrodes. Weighing TF-B520 activated carbon and MWCNTs and mix those two materials into a clean and dry vessel with a mass ratio 2:1 (TF-B520: MWCNTs). Suspended into butanol at a dispersion concentration of 600mg/L; Sonicate for 6 hours at a room temperature. Take 100ml of this hybrid dispersion, add butanol to 400ml and continue sonication for 30min. Vacuum filtration method was used to sediment the mixed materials and get a bulky paper after 1.5 hours. Further treatment was done by vacuum drying for 24h to remove the butanol and water left inside. Weighing the final bulky paper and cut it into small pieces (6mm). The total amount is 58.9mg and each electrode is 4mg.

The electrochemical measurements were performed in a 2025 type coin cell configuration with two symmetric electrodes with same mass loading and same area, LiPF6 in PC was used as electrolyte and glass microfiber filter as separator. Cyclic Voltammetry and Constant Current Charge/Discharge were performed in Arbin 2000.

3. Results and discussion

Cyclic Voltammetry testing on super-capacitor cells assembled with TF-B520/MWCNTs was done to analysis it's electrochemical performance. Figure 1 shows the CV test results of the cell assembled with TF-B520/MWCNTs. Where the Capacitance was calculated according Eq.1:

$$C = \frac{Q}{Um} = \frac{It}{Um} = \frac{I}{(U/t)m} = \frac{I}{vm} \tag{1}$$

Where I is the test current (A), v is the scan rate (A/s), m is the mass of one electrode (g).

At a low scan rate of 0.001V/s, 0.002V/s, 0.005V/s, CV curves assume a rectangular shape, characteristic of good double layer performance. When the scan rate is 0.02V/s, CV curve is non-rectangular, which means the series resistance in cell behaviour becomes more obvious. At a scan rate 0.05V/s and above, the super-capacitor cell shows a completely resistive behaviour.

The difference behaviour of the cell's specific capacitance assembled with TF-B520/MWCNTs at different scan rates is summarized in Figure 2, where the specific capacitance were calculated by Eq. 2, from the integral of discharge half cycles of CV curves.[10]

$$C_{sp} = 2\frac{\int IdV}{vVm} \tag{2}$$

The specific capacitance decreased to 66.4% when the scan rate increased from 1mv/s to 10mv/s. At a scan rate of 20mv/s, its 48.5%, when the scan rate is 0.1V/s, it's only 8.3%. The specific capacitance is almost zero at a rate above 0.2V/s.

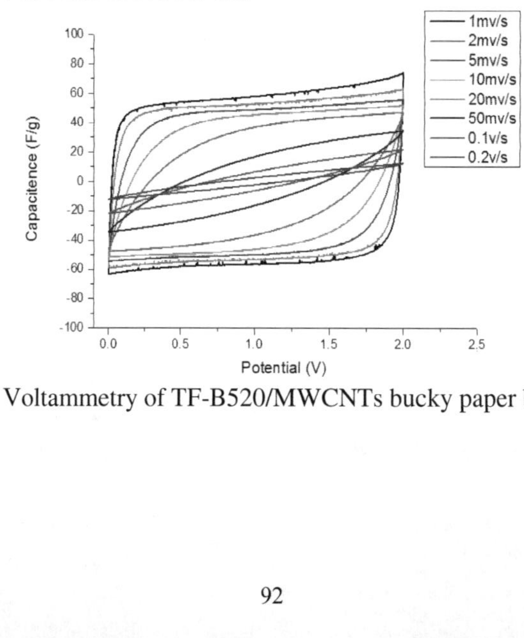

Figure 1. Cyclic Voltammetry of TF-B520/MWCNTs bucky paper based capacitor cell

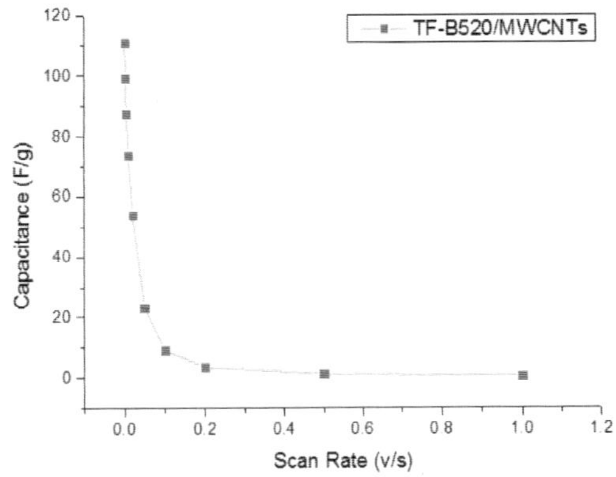

Figure 2. Specific capacitance of TF-B520 electrode in different scan rate

Constant Current Charge/Discharge (CCCD) testing is the most widely used method for the characterization of Super-capacitors under direct current. The vertical drop in voltage at the beginning of the discharge curve (IR drop) shows the cell' resistive behaviour. The CCCD test of SCs assembled with TF-B520/MWCNTs is showed in Figure 3. At low current densities (0.025,0.05A/g), the potential drop is < 0.1V, the electrodes exhibit a low resistance; at a current density of 0.25A/g, the IR drop is 0.25A; at 0.5A/g the drop increases to more than 0.5V that the total capacitance is negatively affected.

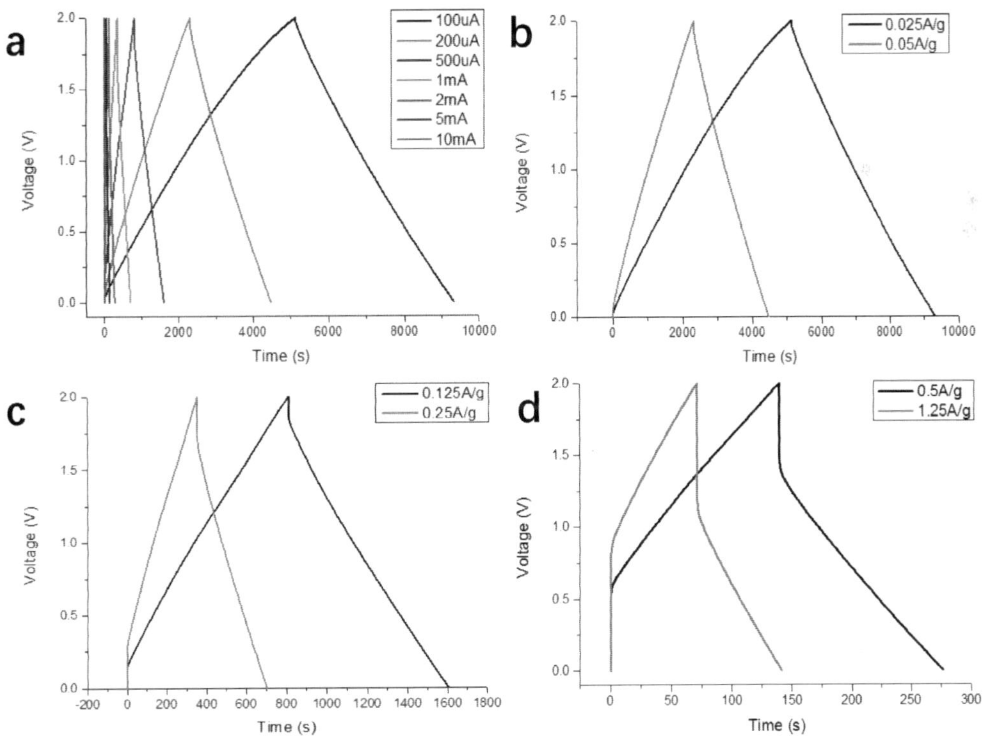

Figure 3. Constant current Charge/Discharge test of TF-B520/MWCNTs electrode in different current

The specific capacitances of one electrode at different current densities were calculated according to Eq. 3, using the discharge curve.[11]

$$C_{SP} = \frac{I\Delta t}{\Delta Vm} \quad (3)$$

Where C_{sp} is the specific capacitance of one electrode (F/g); I is the current of the discharge curve(A); m is the mass of one electrode(g); $\Delta V = V_{max} - V_{IR\,drop}$(V).The specific capacitance decreased fast to 38.6% when the current density is up to 2.5A/g.

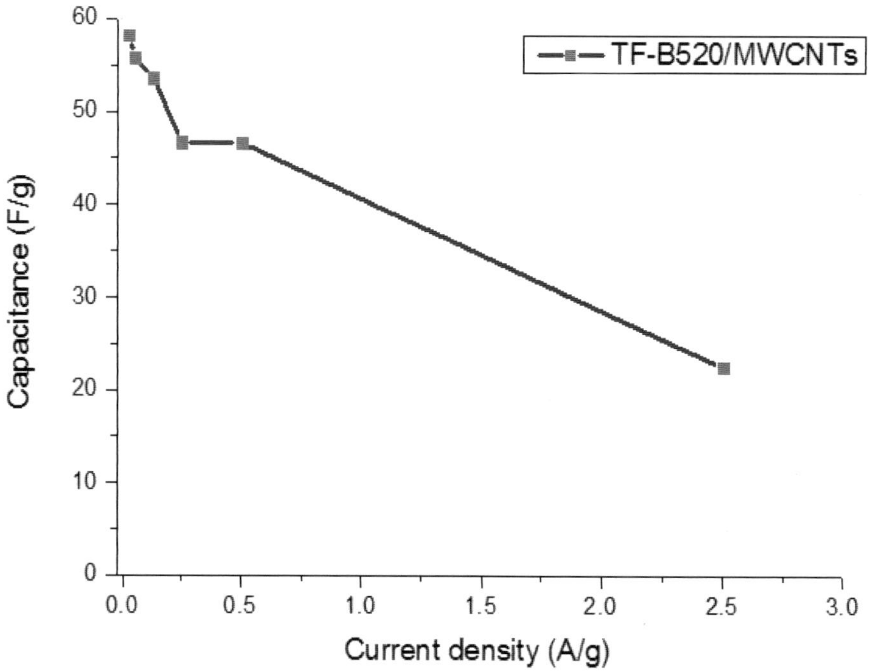

Figure 4. Specific capacitance of TF-B520 cell in different current

The difference behaviour of the cell's specific capacitance assembled with TF-B520/MWCNTs at different current density is summarized in Figure 4, showing that at low current of 0.025A/g, the EDLC cell has the highest specific capacitance of 60F/g, when the current density increase, the specific capacitance will decrease, at high current density 2.5A/g, the cell capacitance is 23F/g.

4. Conclusion

Activated carbon (TF-B520) bulky paper electrode was successfully assembled, and electrochemical behaviour of the electrode was studied by Cyclic Voltammetry and Constant Current Charge/Discharge. The results show that TF-B520 activated carbon bulky paper electrode has a high capacitance of 120F/g at 1mV/s, good capacitor behaviour at scan rate below 20mV/s and current density below 250mA/g. This work shows that the activated carbon TF-B520 can be a good candidate for commercial super-capacitor.

References

[1] C. Largeot, C. Portet, J. Chmiola, P. Taberna, Y. Gogotsi and P. Simon, J. Am. Chem. Soc., 2008, 130, 2730.
[2] S. Kandalkar, D. Dhawale, C. Kim and C. Lokhande, Synth. Met., 2010, 160, 1299.
[3] P. Simon and Y. Gogotsi, Nat. Mater., 2008, 7, 845.
[4] R. Ko¨ tz, S. Mu¨ ller, M. Ba¨ rtschi, B. Schnyder, P. Dietrich, F. N. Bu¨ chi, A. Tsukada, G. G. Scherer, P. Rodatz, O. Garcia, P. Barrade, V. Hermann and R. Gallay, Electrochem. Soc. Proc., 2001, 21, 564.

[5] B. E. Conway, Electrochemical Supercapacitors, Kluwer Academic, Plenum Press, New York, 1999
[6] F. Béguin, V. Presser, A. Balducci, and E. Frackowiak, Adv. Mater., 2014, 26 (14), 2283-2283
[7] A. S. Arico, P. Bruce, B. Scrosati, J.-M. Tarascon and W. van Schalkwijk, Nat. Mater., 2005, 4, 366.
[8] Chmiola, J., Largeot, C., Taberna, P. L., Simon, P. & Gogotsi, Y. Monolithic carbide-derived carbon films for micro-supercapacitors. Science, 2010, 328, 480–483.
[9] Taberna, P. L., Portet, C. & Simon, P. Electrode surface treatment and electrochemical impedance spectroscopy study on carbon/carbon supercapacitors. Appl. Phys. A, 2006, 82, 639–646.
[10] Chen, Y., Zhang, X., Zhang, D., Yu, P. & Ma, Y. High performance supercapacitors based on reduced graphene oxide in aqueous and ionic liquid electrolytes. Carbon 2011, 49, 573–580.
[11] K. Brezesinski , J. Wang , J. Haetge , C. Reitz , S. O. Steinmueller , S. H. Tolbert , B. M. Samarsly , B. Dunn , T. Brezesinski , J. Am. Chem. Soc. 2010 , 132 , 6982 .

Integration of Missing Parameters into a DES Tool

Qian Wang*, **Reda Nujoom and Maji Abubakar**

School of Mechanical and Design Engineering, University of Portsmouth, Portsmouth PO1 3DJ, UK

*E-mail: qian.wang@port.ac.uk

Abstract. Products are usually made by accomplishing a series of manufacturing processes in a sequential flow line that is also known as a manufacturing system. Today, lean methods are widely adopted by many manufacturing plants as a popular model in designing, implementing, operating or managing a manufacturing system. It has been proved as a cost-effective approach to boost system efficiency and productivity by consistently seeking and removing any non-value added activities (i.e., wastes) during a production with a small or without any additional investment. Nevertheless, identification of these wastes using the traditional lean methods does not include such wastes as amounts of energy consumption and CO_2 emissions. For human centered assembly lines, for instance, it is reported that applying highly skilled, flexible and dynamic workers into production lines is also a good practice for implementing a lean manufacturing system in which each worker performs multiple tasks amongst stations. On the other hand, most studies on manufacturing systems using the modelling simulation methods failed to consider parameters of energy consumption, CO_2 emissions and human factors that may also impact the overall system performance. The simultaneous prediction, which relates to amounts of energy consumption and CO_2 emissions and effects of human factors (or human performance) for a manufacturing system evaluation, is often overlooked by researchers or system designers partially due to a lack of existing DES (discrete event simulation) tools that enable incorporating these parameters into an established DES model. This paper presents a study by addressing these issues aiming to incorporate these missing parameters of energy consumption, CO_2 emissions and human factors (age and experience) into a DES model.

1. Introduction

The concept of lean production or manufacturing emphasizes the importance of eliminating "any non-value-added wastes" in every aspect of manufacturing-related activities thereby increasing manufacturing efficiency and productivity at a workplace, reducing time required for manufacturing a product and improving quality of final products. Lean manufacturing can be implemented with lean thinking, which is described as an enterprise culture by recognizing that there is always room for improvement of product design, manufacturing processes or operations, production systems, and management. Thus, any creation of ideas or approaches, which can be utilized effectively and economically for enhancing efficiency and productivity through reduction of system wastes and increase of system responsiveness and flexibility, need to be encouraged. These ideas or approaches must also be embedded in any manufacturing-related activities or organizations. At a typical small-medium enterprise (SME) in manufacturing, multifunctional workers are the key in success of implementing a lean manufacturing system, particularly when the manufacturing system involves a great deal of human-centered operations. Such a manpower production line need also to be designed towards a reduction of the seven wastes (identified by the traditional lean methods), which are the waste of overproduction, the waste of waiting for parts to arrive, the waste of conveyance or transport system, the waste in processing or operations, the waste of inventory, the waste of motion and the waste of rework. Nevertheless, these wastes often do not comprise the environmental wastes in terms of such as energy consumption and CO_2 emissions relating to manufacturing activities; and these environmental wastes also add no value to manufactured products. In practice, the environmental considerations may be addressed as constraints or separately or often overlooked during the design and implementation of a manufacturing system. As for human centred manufacturing systems, most studies have focused on the impact of human factors on human performance in linguistic terms, which are not specifically related to manufacturing activities or production systems. In a human-centred manufacturing system, however, production loss can be caused due to varying human performance that is often affected by a variety of human factors interacting in a complex way, such a phenomenon is often under or overestimated or simply neglected in manufacturing systems design, evaluation and implementation [1]. Moreover, DES (discrete even simulation) tools are often used as an aid for manufacturing systems design and evaluation. These tools, however, are developed focusing on conventional operations of systems and operators. And these tools do not provide facilities that allow system designers to combine parameters of such as energy consumption and CO_2 emissions, and human attributes (or human performance) within an investigation of the overall system performance. This is because, for example, in a DES model, the workers are defined and treated as the same as parts, conveyors and processing machines and so on. The application of DES simulation models is therefore restricted to predicting such variables as the required number of workers, their utilisation percentages or shift patterns and routes [2]. This paper reports a latest development aimed at addressing the above issues by attempting to create a user-friendly method incorporating some parameters of energy consumption, CO_2 emissions and a couple of human factors or attributes into an integrated DES model. With this approach, amounts of energy consumption and CO_2 emissions or effect of human factors on system performance can be quantified.

2. Energy consumption and CO_2 emissions relating to manufacturing systems

The traditional lean approaches can only be used for identifying wastes in terms of manufacturing operations through implementation of management rules (such as 5S rules); these methods do not consider wastes of such as energy consumption and CO_2 emissions in relation to operations or processes of a manufacturing system. In a manufacturing system, energy is used for operating machines, illumination systems, air conditioning systems and other relevant supporting equipment such as compressors which supply compressed air to some of these machines. To describe amounts of energy consumption and CO_2 emissions used for these facility mathematically, the following notations are used [3-4]:

m: number of processes in a manufacturing system
n_i: number of machines involved in process i, where $i \in \{1, 2, ..., m\}$
E_i (kWh): energy consumption for a machine involved in process i
E_i^{cond} (kWh): energy consumption of an air conditioning system
E_i^{illum} (kWh): energy consumption of an illumination system
$E_i^{air\ comp}$ (kWh): energy consumption of compressed air needed for a machine involved in process i
TE (kWh): total energy consumption of a manufacturing system
N_i (kw): installed power for a machine involved in process i
R_i (kg/h): manufacturing rate for a machine involved in process i
τ_i (hr): operating time for a machine involved in process i
μ_i (%): efficiency for a machine involved in process i
∂_i (kg): mass of materials transferred from a machine involved in process i
G_i (kg): mass production per month
$¥_i$ (%): waste ratio for a machine involved in process i

$Ȩ_i$ (kWh): energy consumption of air conditioning per month
$\check{E}_i$ (kWh): energy consumption of illumination per month
$\zeta_i^{air\ comp}$ (kWh/m^3): energy consumption per cubic meter of a compressor
υ_i (m^3/h): compressed air used for a machine involved in process i per hour
$\rho_i^{air\ comp}$ (m^3/h): capacity of compressed air in cubic meter per hour of a compressor
$N_i^{air\ comp}$ (kWh): installed power for a compressor
e_i (kg/kWh): amount of CO_2 emissions per kWh released from a machine involved in process i
Te_i (kg/kWh): amount of CO_2 emissions per KWh of a machine, an air conditioning system and an illumination system involved in process i
ω : CO_2 emission factor using different energy sources
Te (kg/kWh): total amount of CO_2 emissions released from a manufacturing system
q_i (kg): mass of materials involved in process i

2.1. Energy consumption

Energy consumption E_i for a machine involved in process i is given by:

$$E_i = \tau_i \times N_i \times n_i \tag{1}$$

And operating time τ_i for a machine involved in process i is calculated below:

$$\tau_i = \frac{q_i}{R_i \times \mu_i} \tag{2}$$

Mass of materials q_i transferred from a machine involved in process i is obtained by:

$$q_i = \partial_i \times (1 + ¥_i) \tag{3}$$

Energy consumption for air conditioning E_i^{cond} in a manufacturing system is given by:

$$E_i^{cond} = Ȩ_i \times \frac{\partial_i}{G_i} \tag{4}$$

Energy used for an illumination system Eiillum is calculated by:

$$E_i^{illum} = \check{E}_i \times \frac{\partial_i}{G_i} \tag{5}$$

Energy consumption of a compressor needed for a machine involved in process i, thus $E_i^{air\ comp}$ is calculated by:

$$E_i^{air\ comp} = \tau_i \times \zeta_i^{air\ comp} \times \upsilon_i \times n_i \tag{6}$$

Where $\zeta_i^{air\ comp}$ can be determined by:

$$\zeta_i^{air\ comp} = \frac{N_i^{air\ comp}}{\rho_i^{air\ comp}} \tag{7}$$

Thus, total energy consumption TE for a manufacturing system is given below:

$$TE = \sum_{i=1}^{m}(E_i + E_i^{air\ comp} + E_i^{cond} + E_i^{illum}) \tag{8}$$

Where, $i \in \{1,2,...., m\}$

Hence, equation 8 can be given below:

$$TE = \sum_{i=1}^{m}\left(\frac{\partial_i \times (1+¥_i)}{R_i \times \mu_i} \times N_i \times n_i + \tau_i \times \zeta_i^{air\ comp} \times \upsilon_i \times n_i + E_{\ell i} \times \frac{\partial_i}{G_i} + \check{E}_i \times \frac{\partial_i}{G_i}\right) \tag{9}$$

2.2. CO_2 emissions

Amount of CO_2 emissions e_i released from a machine involved in process i is calculated by

$$e_i = \omega \times E_i \tag{10}$$

And total amount of CO_2 emissions *Te* can be calculated as follows:

$$Te = \sum_{i=1}^{m}(e_i \times q_i + 0.6895 \times (E_i^{air\ comp} + E_i^{cond} + E_i^{illum})) \tag{11}$$

Where $i \in \{1, 2, ..., m\}$
Thus, Te can be expressed below:

$$Te = \sum_{i=1}^{m}(\omega \times \tau_i \times N_i \times n_i \times \partial_i \times (1+¥_i) + 0.6895 \times (\tau_i \times \zeta_i^{air\ comp} \times \upsilon_i \times n_i + E_{\ell i} \times \frac{\partial_i}{G_i} + \check{E}_i \times \frac{\partial_i}{G_i}) \tag{12}$$

3. Walking workers and human factors

It was reported that approximately one third of all German companies that have invested in highly advanced automaton have recognized that these solutions are not flexible enough and have reduced again their level of automation; 38% of these companies have reduced automation by taking advantage of a more efficient use of their qualified workforce [1]. Figure 1 illustrates a lean manufacturing system using multifunctional, dynamic and cross-trained walking workers. Within such a system, each worker travels with a partially assembled product downstream and stops at each station carrying out the essential assembly work as instructed. Each worker is previously trained to be capable of performing assigned tasks to build a product completely from start to end along the line. This type of system inherently prevents unnecessary in-process inventory thereby decreasing the buffer requirement [5]. For such a human centered manufacturing system, however, the overall system performance can be affected by varying performance of individual workers who have such as different ages and experience. Effects of these human factors or human performance can also be unpredictable and it may alter due to varying psychological and physiological states [6].

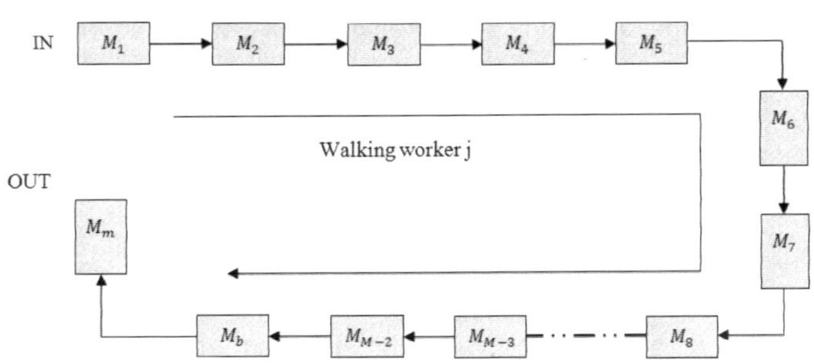

Figure 1. Walking worker production

3.1. Ageing

There are some studies in a view of socio-technical or psychological sciences to evaluate the effects of human factors or human performance relating to the design of manufacturing systems. Nevertheless, these studies are basically described in a form of general language that manufacturing engineers often find difficult to understand. A literature review by authors [7] indicates that that human performance starts to deteriorate at 38 years old as a base line. It also shows the loss rate, which refers to the rate of decline as the remaining functional capacity (%) of his/her peak at the age of 38 using the regression analysis method, can be described below:

$$F_{rm} = k_2 - L_r(k_1 - 38) \tag{13}$$

In which,

$L_r = 0.57 + 0.012\text{k}$

where,
L_r = Loss rate in percentage

k = Age in years

F_{rm} : Remaining capacity in percentage of the peak at 38 years old

k_2: Peak capacity (100%) at 38 years old

L_r: Loss rate in percentage

k_1 : Existing age in years

3.2. Experience

Experience can be defined as the knowledge or a skill to be gained through involvement of a specific task, event or subject. In a human centred assembly line, for instance, experience can be quantified as a learning curve that refers to a trend in reduction of assembly time for an assembly work as quantity of products increases through a learning and forgetting process, which can be denoted as follows [8]:

$$T_n = T_t \cdot Q^c \tag{14}$$

Where
T_n: Average time to produce the nth unit
T_t: Assembly time to produce the first unit
Q: Cumulative number of units produced
C : Learning index which determines the speed of learning occurring each time as a cumulative output increases, it is computed as $\frac{\log(R)}{\log(2)}$ where the learning rate R is measured in percentage ($0 < R < 1$) [9]. Note that the average time towards the steady state decreases with the increase of number of units produced. Thus, the average time towards a steady stage can be given as:

$$T_A = T_t \cdot B^R \tag{15}$$

or

$$T_t = \frac{T_A}{B^R} \tag{16}$$

Where
T_A: The average time towards a steady stage
B: Batch size
it yields:

$$T_n = \frac{T_A}{B^R} \cdot Q^R \tag{17}$$

Hence

$$T_n = T_A(\frac{Q}{B})^R \tag{18}$$

The loss in average assembly time per worker due to ageing is given below:

$$\Delta_{Lt} = T_n \times F_{dl} \tag{19}$$

Where

F_{dl} : Kinematic decline rate (%) of human full capacity

$$\Delta_{Lt} = T_A(\frac{Q}{B})^R \times F_{dl} \tag{20}$$

Hence, average total assembly time per worker associated to ageing is computed below:

$$T_{At} = T_A(\frac{Q}{B})^R + T_A(\frac{Q}{B})^R \times F_{dl} \tag{21}$$

Where

Δ_{Lt}: Average assembly time loss due to ageing

T_{At}: Average total assembly time per worker due to ageing

4. Integration of missing parameters into a DES tool

A feasibility study was carried out by establishing a user-friendly way incorporating identified parameters of energy consumption, CO_2 emissions and effects of human factors into a DES tool, which can permit manufacturing system designers, at the early design stage, to evaluate the overall performance of a manufacturing system with considerations of these parameters. These parameters can be modeled into either external MS Excel worksheets that can be actively linked into a DES tool or the internal program using the DES simulation language or other language like Java. These parameters interact with parameters of physical elements (built in the DES tool) of machines and conveyors etc., together with logical interrelationship for operational activities in a manufacturing system. Thus, this method enables system designers to evaluate the overall performance of a manufacturing system considering not just operational activities but also amounts of energy consumption and CO_2 emissions. This integrated modeling simulation methodology permits users to determine the relevant impact on logical interactions and interrelationships between parameters in manufacturing operations, energy consumption, CO_2 emissions, and human factors within the created manufacturing system model. Figure 2 illustrates the connection interface between the DES model and input/output Excel files in which both input data and output data (simulation results) can be manipulated in an Excel environment.

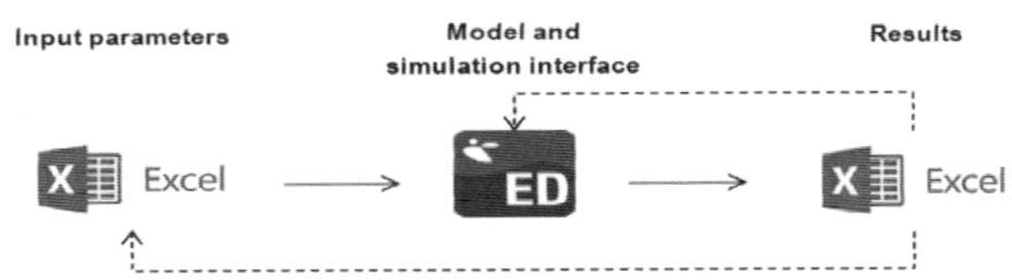

Figure 2. Connection between the DES model interfaced with input/output Excel files

Figure 3 shows the simulation results in terms of amount of CO_2 emissions of each processing task using the alternative energy sources, which are subject to the selection of CO_2 emission factors. It can be seen that the highest amount of CO_2 emissions is 4.08×10^9 kg using oil as indirect energy source, 2.9×10^9 kg using oil as direct energy source and 3×10^6 kg using solar as indirect energy source to generate electricity, respectively. The results shown in Figure 3 also indicate that the total amount of CO_2 emissions using the solar energy is 7×10^6 kg, which is the lowest, followed by 6.5×10^9 kg using oil as direct energy source and 8.4×10^9 kg using oil as indirect energy source to generate electricity. Figure 4 shows the trend in decline of human functional capacity at varying ages after 38 years old. Figure 5 shows the trend of average assembly time corresponding to accumulative number of units produced by workers at the age of 38 years old during a learning process of repetitive operations of assembling a unit. It can be seen that the average assembly time tends to be stabilised after performing assembly of over 480 units.

IOP Conf. Series: Materials Science and Engineering **436** (2018) 012016 doi:10.1088/1757-899X/436/1/012016

5. Conclusion

The paper reports a feasibility study by proposing an integrated lean approach aimed at not merely improving efficiency and productivity of a manufacturing system by reducing unnecessary production wastes, which can be identified using the traditional lean methods, but also examining wastes of energy consumption and CO_2 emissions within a manufacturing system. For human centred manufacturing systems, effects of human factors can also be examined using the DES tools. This is because current DES tools in the market do not have functionality allowing manufacturing systems designers to create a DES model that considers parameters relating to energy consumption, CO_2 emissions and effects of human factors or human performance. The paper presents an effective and user-friendly method incorporating some parameters relating to energy consumption, CO_2 emissions and effects of human factors to a DES model. The simulation results demonstrate the applicability of using this method by quantifying the amount of energy consumption, CO_2 emissions or effects of human factors on the overall system performance.

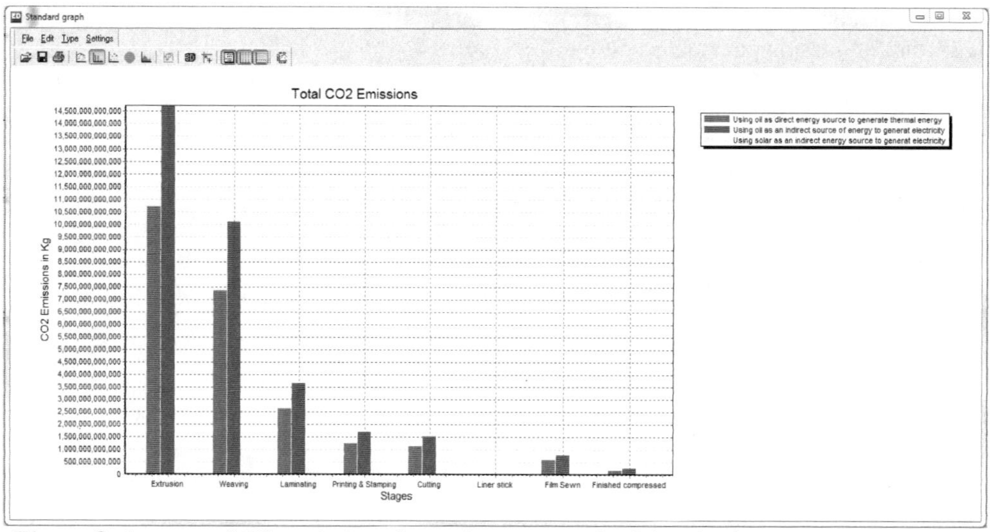

Figure 3. Amount of CO_2 emissions for each process task of the plastic woven-sacks production using alternative energy sources.

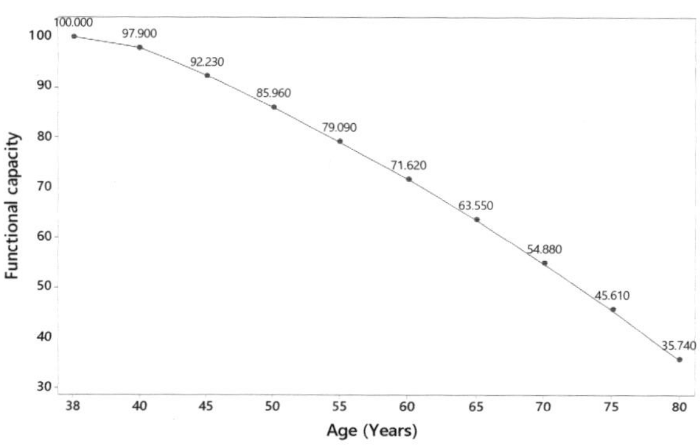

Figure 4. Decline in human functional capacity after 38 years old

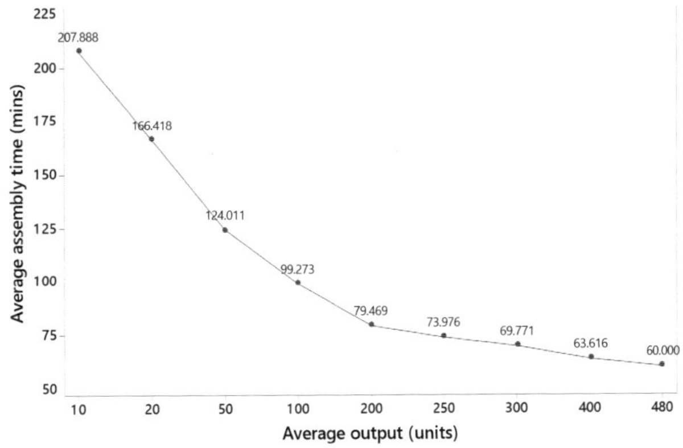

Figure 5. Average assembly time (mins) vs accumulated output of workers at 38 years old

Acknowledgments
The authors would like to thank the industrial participants for their contributions of a survey to this project.

References

[1] Bley H, Reninhart G, Seliger G, Bernardi M and Korne T 2004 Appropriate human involvement in assembly and disassembly *CIRP Annals*, vol. 53 p 487–509
[2] Wang Q and Chatwin C R 2005 Key issues and developments in modelling and simulation-based methodologies for manufacturing system analysis, design and performance evaluation *International Journal of Advanced Manufacturing Technology* Vol. 25, p 1254-1265
[3] Nujoom R, Wang Q and Mohammed A 2018 Optimisation of a sustainable manufacturing system design using the multi-objective approach, *International Journal of Advanced Manufacturing Technology*, Vol.96 p 2539-2558
[4] Nujoom R, Mohammed A and Wang Q 2017 A sustainable manufacturing system design: a fuzzy multi-objective optimization model *Environmental Science and Pollution Research*, in press
[5] Wang Q, Lassalle S, Mileham A and Owen G 2009 Analysis of a linear walking worker line using a combination of computer simulation and mathematical modeling approaches *Journal of Manufacturing Systems* vol. 28 p 64–70
[6] Wang Q, Sowden M and Mileham A R 2012 Modelling human performance within an automotive engine assembly line *International Journal of Advanced Manufacturing Technology* vol. 68(1-4) p 141-148
[7] Abubakar M I and Wang Q Key human factors and their effects on human centered assembly performance *International Journal of Industrial Ergonomics* under second review
[8] Jaber M Y 2001 Learning curves theory, models and applications. London: CRC Press
[9] Anzanello M J and Fogliatto F S 2011 Learning curve models and applications: literature review and research directions. *International Journal of Industrial Ergonomics* p 573-583

Research on Seawater Corrosion Resistance of Spray Polyurea Protective Coating

Lyu Ping[1,*], Lin Jing[1], Ma Mingliang[1] and Song Yilong[1]

[1]Qingdao University of Technology, College of Civil Engineering, 11 Fushun Road, Qingdao 266033, China

*lyuping-qut@sina.com

Abstract. In this paper, the changes of mechanical properties and surface properties of QF-162 polyurea coating after soaking in seawater for 120 days were studied. The changes of internal chemical bonds before and after coating corrosion were observed with Fourier Transform infrared spectroscopy (FT-IR). The results of mechanical properties study showed that after soaking in seawater for 120 days, the tensile strength of QF-162 polyurea coating decreased by 4.23%, the elongation at break decreased by 7.10%. The hardness changed slightly and Shore A hardness decreased by 3.23%. Surface performance results showed that the contact angle of QF-162 polyurea coating decreased with the increase of soaking time, and the surface energy showed a trend of increase. After soaking 120 days, the contact angle of polyurea coating decreased by 6.08%, surface energy increased by 15.83%. FT-IR microscopic results showed that there was no obvious fracture phenomenon in the internal chemical bonds of QF-162 polyurea coating and the coating had good anti-corrosion to seawater soaking.

1. Introduction

At present, with the rapid development of marine engineering construction in China, more and more offshore buildings (such as cross-sea bridges, undersea tunnels) have appeared [1-2].Affected by the bad marine environment, many marine buildings appeared corrosion and exfoliation, even collapse phenomenon [3-5]. The durability of building structures in marine environment had become the focus of attention of many experts and scholars [6-9]. Coating protection technology was a simple and effective method to protect concrete building structure. It was particularly important to study the performance of the marine environmental building structure protective coating, to grasp the variation rule of the performance of the protective coating in the marine environment, and to improve the service life of the building structure [10].

Polyurea was a new environment-friendly material which did not contain a catalyst, and its curing speed was fast. It could be sprayed on any surface, vertical plane and inclined plane. It had excellent corrosion and aging resistance. In this paper, the QF-162 polyurea coating developed by the institute of functional Materials of Qingdao University of Technology was selected to carry out the seawater corrosion resistance experiment, and the variation rule of the seawater corrosion resistance of the QF-162 polyurea coating after soaking in seawater for 120 days was discussed.

2. Experiment

2.1 Sample preparation

The polyurea material used in this paper was made by the laboratory of Qingdao University of Technology. It was coated with a thickness of about 2.0mm using PMC spraying equipment in the United States. The materials required for tensile test were cut into dumbbell shape by slicing machine.

2.2 Experimental equipment and test methods

Tensile testing: the slicer (MZ-4102) and universal testing machine (MZ-4000D_1) used in the experiment are all provided by Jiangsu Ming Zhu experimental machinery co., LTD. According to GB/T 528-2009, the coating was cut into dumb bell shape for testing.

Hardness test: adopt the rubber and plastic bench measurement meter provided by Jiangsu Ming Zhu experimental machinery co., LTD. According to the provisions of GB/T 531.1-2008, after the superposition and leveling of three layers of samples, the measurement was carried out by using Shore hardness tester.

Contact angle test: the SDC-200 contact angle measuring instrument provided by Dongguan Sheng Ding Precision Instrument Co., Ltd., Using a syringe with a needle, a drop of ultrapure water droplet was placed on the surface of the sample to be tested, and stopped for a few seconds. The contact angle was measured within 30 seconds after the drop was stabilized. Finally, the arithmetic average of five points was chosen to represent the final result.

FT-IR test: the Fourier infrared spectrometer (Nicolet 8700) provided by American Thermo Nicolet Corporation, with a resolution of 4cm-1and a spectral range of 7800~350cm^{-1}.

2.3 Experimental environment

The sea water pool on the roof of the No. 2 experimental building of Qingdao University of Technology was selected as the experimental site for seawater corrosion. According to the requirements of GB/T 1690-2010, the prepared samples were placed in the sea basin for soaking experiments, and the changes of mechanical properties (tensile strength, fracture elongation and Shore A hardness) and surface properties (contact angle and surface energy) were measured respectively when soaked for 30 days, 60 days and 120 days.

3. Results and discussion

3.1 Mechanical properties

Figure 1 showed the variation of tensile strength and elongation at break of QF-162 polyurea coating after soaking in seawater for 30 days, 60 days and 120 days respectively. The Shore A hardness of QF-162 polyurea coating after soaking in seawater for 120 days was shown in figure 2.

As can be seen from the curve in the figure 1, the tensile strength and elongation at break of polyurea coating generally showed a downward trend after being soaked in seawater for 120 days. After soaking for 120 days, the tensile strength decreased from 22.68MPa to 21.72MPa, decreased by 4.23%. The elongation at break reached 473.52% after 120 days, decreasing by 7.10%. The tensile strength and elongation at break of polyurea coating remained above 20MPa and 400% after soaking for 120d, respectively. This indicated that seawater had little influence on the tensile properties of QF-162 polyurea coating, and the coating had good elastic properties.

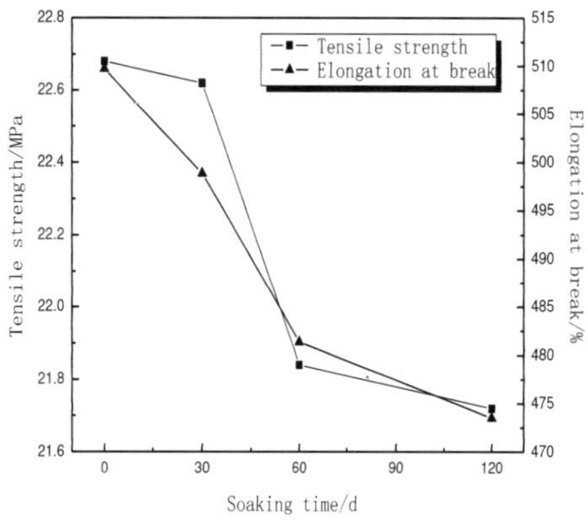

Figure 1. Variation curve of tensile properties of QF-162 polyurea coating after soaking in seawater for 120 days.

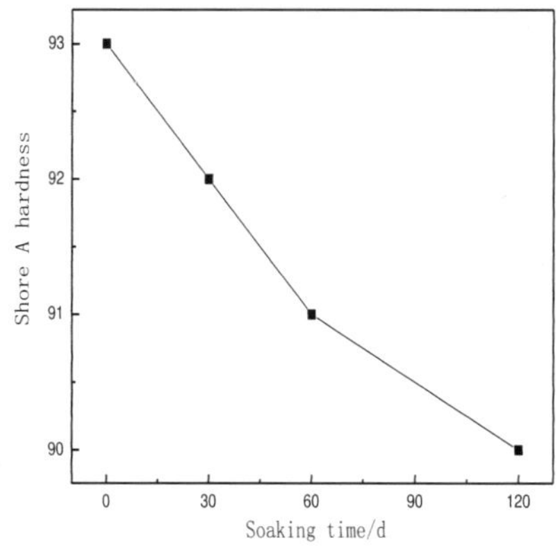

Figure 2. Changes of Shore A hardness of QF-162 polyurea coating after soaking in seawater for 120 days.

As can be seen from figure 2, after soaking in seawater for 120 days, the Shore A hardness of QF-162 polyurea coating generally showed a downward trend, but the change was not significant. After soaking for 120 days, the Shore A hardness of polyurea coating decreased by 3.23%, but remained around 90.

It can be seen from figure1 and figure2, seawater soaking corrosion had a certain effect on QF-162 polyurea coating, but the overall effect was small. The polyurea coating had good anti-corrosion to seawater. After soaking in seawater for 120 days, the tensile strength remained above 20MPa. Shore A hardness remained around 90. The high tensile strength and hardness indicated that the molecular structure of polyurea coating was compact and no swelling damage occurs. The infiltration of the coating was further analyzed by studying the surface performance.

3.2 Surface performance

The change of contact angle and surface energy of QF-162 polyurea coating at 120 days was tested. It reflected the change of wettability of polyurea coating before and after seawater immersion. The experimental results were shown in figure 3.

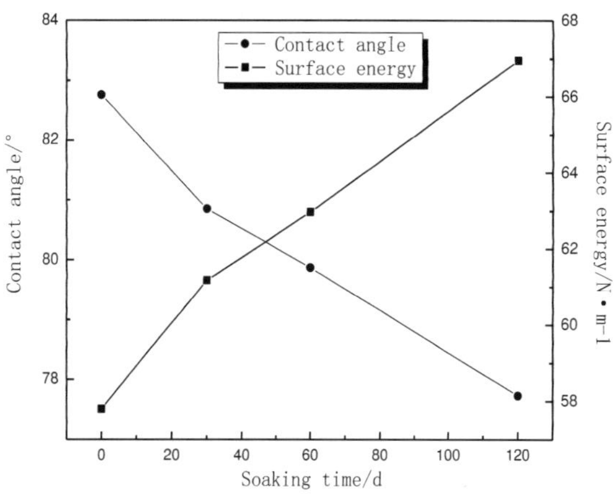

Figure 3. Changes in surface properties of QF-162 polyurea coating with soaking time.

Figure 3 showed the change of contact angle and surface energy of QF-162 polyurea coating with seawater soaking time. The experimental results showed that the contact angle of QF-162 polyurea coating decreased with the increase of soaking time, while the surface energy increased. This indicated that the wettability of polyurea coating increased with the increase of seawater soaking time. Before seawater soaking, the contact angle of polyurea coating was 82.76 °and the surface energy was 57.80 $N \cdot m^{-1}$. After soaking for 30 days, 60 days and 120 days, contact angle decreased by 2.31%, 3.49% and 6.08% respectively. The surface energy increases by 5.85% and 15.83% respectively at the above time points. The reason for the change in contact angle and surface energy of polyurea coating might be due to the uneven change of coating surface affected by corrosion ions in seawater. The corrosion ions directly contacted with the surface of the coating or infiltrate into the coating through micropores, which destroyed the coating and enhance the wettability of the coating. The change of coating internal structure would be further analyzed by FT-IR spectrum.

3.3 FT-IR analysis

The FT-IR spectra of QF-162 polyurea coating after soaking in seawater for 30 days, 60 days and 120 days were shown in figure 4. As can be seen from figure 4, near 3360cm^{-1} was the N-H stretching vibration peak. The absorption peak was relatively strong, and it was a typical stretching vibration absorption peak of hydrogen bonding N-H. That means that N-H on the urea group was almost completely hydrogen bonded. Near 1530cm^{-1} was the stretching vibration peak of C-N and N-R. Within the range of 2965~2870cm^{-1} was the C-H stretching vibration peak. The characteristic peak of C=O was within the range of 1600~1700cm^{-1}. The 1100~1016cm^{-1} range was the telescopic vibration peak of the C-O-C. The existence of the above characteristic peaks proved that the materials contain urea bonding -NHCONH-. The comparison of four curves a, b, c and d in figure 4 can be found, the FT-IR spectra of polyurea coating were basically consistent after being soaked for 30 days,60 days and 120 days. It indicated that no chemical reactions such as chemical bonds breaking or redox occurring in the coating.

A combination of figure 1 and figure 4 can be found, QF -162 polyurea coating was soaked in seawater for 30 days, 60 days and 120 days, the tensile strength decreasing, the elongation at break decreasing slightly, and the hardness did not change much. But the internal structure of the material had not change in any form. That is, during the whole process of seawater immersion, the polyurea coating had not suffer any structural damage. It showed that seawater immersion corrosion had little effect on the internal structure of polyurea coating.

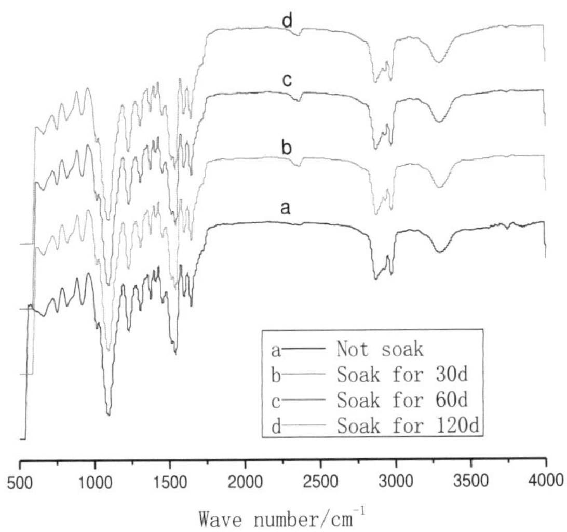

Figure 4. FT-IR spectra of QF-162 polyurea coating before and after seawater soaking.

4. Conclusion

It can be seen from the above experimental results, the QF-162 polyurea coating maintained good corrosion resistance in 120 days seawater immersion corrosion environment. Tensile strength and elongation at break showed a general downward trend. The tensile strength reached the lowest when soaked for 60 days, which was 21.72MPa. The elongation at break remained above 400% after soaking 120 days, indicating that the QF-162 polyurea coating had good elastic properties. FT-IR results showed that the internal structure of polyurea coating did not change. It showed that polyurea coating had excellent anti-corrosion.

References

[1] Itoh, Y, Gu, HS. Prediction of Aging Characteristics in Natural Rubber Bearings Used in Bridges [J]. Journal of Bridge Engineering.2009, 14 (2):122-128.
[2] H.S. Gu, Y. Itoh. Aging Behaviour of Natural Rubber and High Damping Rubber Materials Used in Bridge Rubber Bearings[J]. Advances in Structural Engineering.2010, 13(6): 1105-1113.
[3] Huang Weibo. Spray polyurea elastomer technology [M]. Chemical Industry Press.2005.
[4] Huang Weibo, Liu Xudong, Lv Ping, et al. Key Point Analysis of Beijing-Shanghai High Speed Railway Polyurea Protective Project [J]. Journal of Modern Paint and Finishing.2012,15(8): 11-15.
[5] Lv Ping, Shi Shifan, Xiang Jiayu, et al. Spray pure polyurea technology applied to improving the durability of concrete cross- sea bridge [J]. Journal of Concrete.2012, (8):119-121.
[6] Lv Ping, Li Zhigao, Zhang Jing. Anti- corrosion Behaviors of Polyurea Coating in Acid, Alkali and Salt Solution[J]. Journal of Corrosion and Protection.2011, 32(2):103-106.
[7] Huang Weibo, Hu Xiao, Xu Fei. Research progress of protection technologies for scour-abrasion resistance of hydraulic concrete [J]. Journal of Water Resources and Hydropower Engineering.2014, 45(2): 61-63.
[8] Huang Weibo, Xiang Jiayu, Liu Xudong. Polyurea Technology Applied to Post-earthquake Disasters for Defencing [J]. Journal of Earthquake Resistant Engineering and Retrofitting. 2012,34(2):115-118.
[9] Huang Weibo, Xiang Jiayu, Jiang Linlin, et al. Study on pure polyurea protective technology and applications of hydraulic engineering [J].Journal of New Building Materials.2012,39(4):7-9.
[10] Almusallam A A, Khan FM, Dulaijan SU, et al. Effectiveness of sur-face coatings in improving concrete durability[J].Cement and Concrete Composites.2003, 25 (4) : 473-481.

Microstructural Development and Mechanical Properties of Multi-pass Cold Drawn Hypereutectoid

Z H Chen[1], M Z Wei[1], X M Zhang[1], D P Wang[1], K Sun[1], J Zhou[1,2,*] and F Xue[1, 2]

[1] Jiangsu Key Laboratory for Advanced Metallic Materials, Southeast University, Nanjing 211189, China
[2] Jiangsu Key Laboratory of Advanced Structural Materials and Application Technology, Nanjing Institute of Technology, Nanjing 211189, China

*jethro@seu.edu.cn.

Abstract. In the present work, processes of multi-pass cold drawn were conducted on hypereutectoid pearlite steel wires with carbon content of 1.02 w.t.%. Relationship between microstructural development and mechanical properties of hypereutectoid steel wires after drawing was investigated. It was found that when the strain was relatively small, lamellar morphology of the pearlite in the steel wire after cold drawn remained, and the spacing of lamella was refined. With the increase of the compression ratio, coordinated deformation of ferrite and cementite occurred and pearlite lamella were bended and folded. As the strain increased, fibrous morphology of pearlite lamella was observed in longitudinal section of the wire. High magnification TEM observation revealed the fragmentation of cementite and a large number of nano-crystalline. Tensile strength, microhardness and torsion strength of the steel wires after multi-pass drawn process increased significantly with the increase of accumulated strain. The results of XRD analysis show that the <110> peak of the ferrite shifted to the left and widened following the multi-pass cold drawn deformation, indicating that the lattice constant of ferrite was increased. The increase of lattice constant of ferrite was the result of the distortion of ferrite lattice due to the dissolution of carbon atoms.

1. Introduction

Owing to its abundant resources, low cost, excellent mechanical performance, high strength and good ductility, hypereutectoid steel wires have been widely used in piano string[1], automobile tires[2] and suspension bridge cable[3].

To further improve the mechanical properties of hypereutectoid steel wires, much effort has been made to investigate the microstructural evolution and its effect on the mechanical property of the wires. The strength of cold drawn steel wires can be effectively increased by the refinement of interlamellar spacing, the increment of carbon content, the addition of alloying elements, and the increase of deformation strain[4-5].

It was shown that with the drawing process going, the orientation of ferrite and cementite platelets turned to drawn axial gradually, followed by the decrease of pearlite spacing[6]. Finally, the pearlite lamellar will be refined as the fibrous morphology[7]. In addition to pearlite lamellar refinement, dislocation multiplication in ferrite was also proposed by Lenasson et al.[8].

Pearlite is a two-phased, lamellar structure composed of alternating layers of ferrite and cementite. It has been recently emerged revealing that cementite lamellae also exhibited remarkable ductile behavior during drawing[9]. In the process of drawing, in addition to the changes in the spacing between the pearlite laminates, there will also be some coordination deformation between ferrite and cementite[10].

It was found that cementite platelets were divided into nanometer-sized grains as a result of the dissolution of cementite after the serious deformation of wires[11-12]. Under further drawing, the amorphous cementite will be found and distributed in the carbon-rich region between ferrite grains[13-14].

In this article, the changes of microstructure and mechanical properties of multi-channel pull-out hypereutectoid steel wire were analyzed. And the relationship between cementite recrystallization and mechanical properties of the hypereutectoid steel wires were discussed.

2. Experimental methods

Materials used in this study were hot-rolled hypereutectoid steel rods with a composition in the mass fraction (%) of Fe-1.02C-0.297Cr-0.215Si-0.31Mn-0.0041S-0.0089P. After pickling and phosphating, the rods (5.5 mm in diameter) were cold drawn to 1.35 mm in diameter by multi-pass drawing process (the total reduction of 75.5 %, ε=2.38). The temperature of hypereutectoid steel wires during drawing was below 100 °C.

The torsion tests of steel wires were performed on CTT500 torsion tester at room temperature with a rotating speed of 720 °/min. The length of the test wires is 50 times of the wire diameter. Both the torque and torsional angle were recorded in real time. Five identical samples were tested under each condition.

For microstructural observation, after standard grinding, polishing and electro-polishing were applied to prepare samples, the specimens were etched by 4% nitric acid alcohol solution and examined using FEI Siron-400 scanning electron microscope (SEM).

TEM samples were prepared by GATAN 691 Precision Ion Polishing System (PIPS) after the thickness of the samples was mechanically thinned to below 40 μm. Sample was cooled by liquid nitrogen in PIPS to avoid the temperature rise. Ion angle and energy were set to 3° and 2 Kev at the final stage of TEM sample preparation to minimize microstructure damage.

3. Results

3.1 Microstructure of hypereutectoid steel wires

Fig. 1 shows the microstructures of the cross-section (a and b) and longitudinal-section (c and d) of the hypereutectoid steel wire. The strain is 0, 1.34, 0, 1.34 in sequence. In fig. 1 a and c, it can be seen that the lamellar structure of hypereutectoid steel wire consists of the white part cementite and the black part ferrite, who stay parallel to each other. And the grain orientations of the pearlite are homogeneous. The selected areas in fig. 1 marked by the red circular line show the microstructure evolution of the lamellar structure. It can be seen from fig. 1 b and d that with the increase of the strain, the pearlite interlaminar spacing decreases gradually, and the ferrite and cementite coordinate deform violently. Additionally, the lamellar structure is bent or broken into pieces for those who perpendicular to the drawing direction as can be seen in fig. 1 b and d, while it shows the fibrous morphology for those who parallel to the drawing direction as can be seen in fig. 1 d.

Fig. 2 presents TEM micrographs of pearlite morphology. The strain is 0, 1.34, 0, 1.34 in sequence. Fig. 2 a and b are the bright-field image of the hypereutectoid steel wires, of which white part is cementite and the black part is ferrite, show the dislocation arrangement after large strain. A number of high density dislocation cells were observed in ferrite lamellae. It is worthy to be mentioned that, not only the ferrite lamellae, but also the cementite lamellae have a reduction in thickness with the increase of the strain. In addition, it has a different reduction level of thickness in different areas. Fig. 2 c and d are the dark-field image of the hypereutectoid steel wires, of which present the evolution of fractured cementite. It can be seen that a number of nano-crystal cementite is distributed in cementite lamellae unevenly. Meanwhile, cementite lamellae were bent due to stress concentration caused by these dislocations. It can be observed in fig. 2 d that the different deformation degree of the cementite lamellae is associated to the amount of fractured cementite. As is marked in the red box line, the area which has the high density fractured cementite has the lower deformation degree.

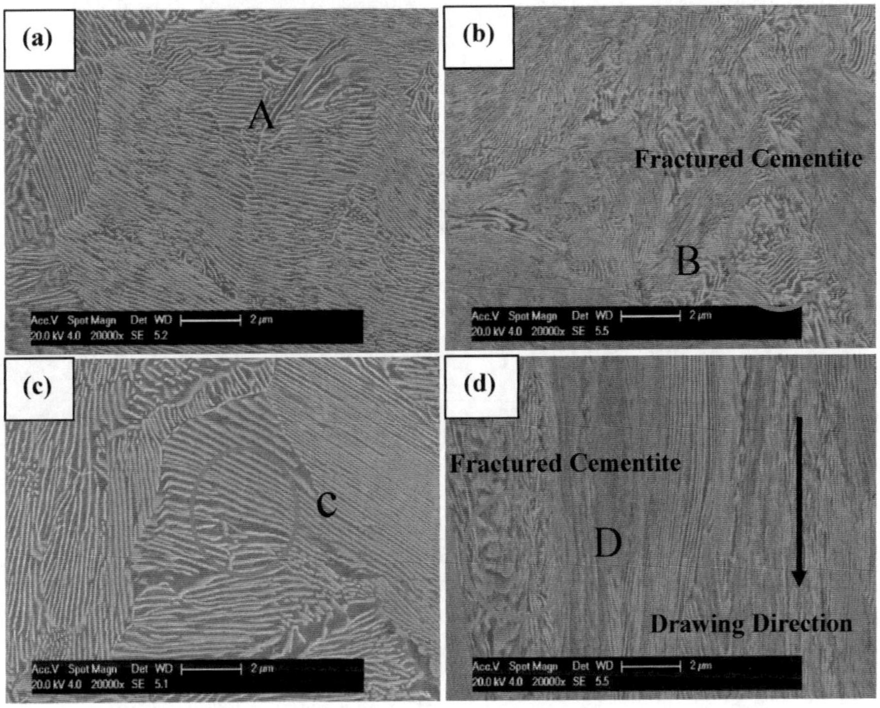

Figure 1. SEM images of hypereutectoid steel wires. (a) and (b) The cross-section morphology. (c) and (d) The longitudinal-section morphology. (strain: a, c=0; b, d=1.34)

3.2 Mechanical properties

Table 1 presents the torsion test results of multi-pass drawing steel wires. Fig. 3 shows the engineering strain-torsional strength/turns curves of the multi-pass cold drawn treated steel wires. It can be seen from table 1 that the strain rates go along with the increase of drawing turns, which results in the reduction of diameter of steel wire. It can be seen from fig. 3 that with the strain increase from 0 to 2.81, torsional strength increases from 1126.83 MPa to 2527.22 MPa. As is shown in fig. 3 the black broken line, the variation trend of torsional strength is increasing, which means the work hardening rate of hypereutectoid steel wires is improved with the increase of strain. It is worthy to be mentioned that the torsional turns firstly increased from 15 to 20 turns and then decreased to 13 turns with the increase of strain, which is different from the direct proportional relationship between engineering strain and torsional strength during the whole multi-pass drawing process.

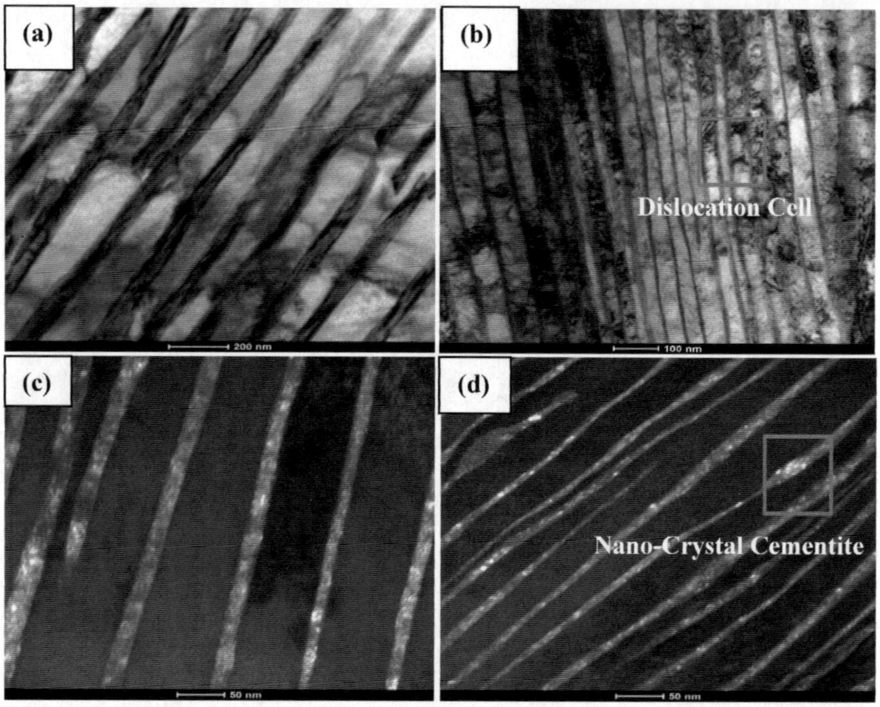

Figure 2. TEM micrographs of cold drawn hypereutectoid steel wires. (a) and (b) Bright-Field image. (c) and (d) Dark-Field image (strain: a, c=0; b, d=1.34)

Table 1. The torsion test results of multi-pass drawing steel wires

Strain	Diameter(mm)	Torsional turns	Torsional strength(MPa)
0	5.5	15	1126.83
0.66	3.96	18	1298.54
1.34	2.81	20	1586.85
2.2	1.83	14	2000.16
2.81	1.35	13	2527.22

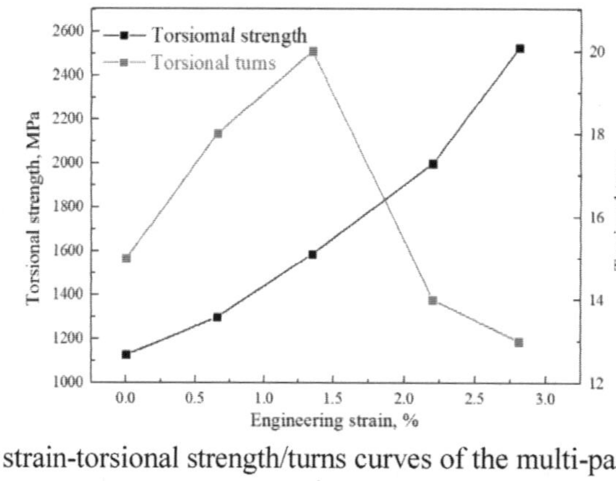

Figure 3. Engineering strain-torsional strength/turns curves of the multi-pass cold drawn treated steel wires

4. Discussion

4.1 Microstructure of hypereutectoid steel wires

It can be seen from fig. 1 and fig. 2 that with the increase of the strain, the pearlite interlaminar spacing decreases gradually, and the ferrite and cementite coordinate deform violently. And the lamellar structure is bent or broken into pieces for those who perpendicular to the drawing direction, while it shows the fibrous morphology for those who parallel to the drawing direction. Meanwhile, a number of high density dislocation cells were observed in ferrite lamellae, while a number of nano-crystal cementite is distributed in cementite lamellae unevenly.

It can be draw from the above experimental results that with the increase of the strain, a number of high density dislocation cells and nano-crystal cementite were formed in ferrite lamellae and cementite lamellae, respectively, which directly resulted in the deformation of pearlite. Those grains who parallel to the drawing direction show the fibrous morphology, while those who perpendicular to the drawing direction is bent or broken into pieces.

4.2 Mechanical properties

It can be seen from fig. 3 that with the strain increase from 0 to 2.81, torsional strength increases from 1126.83 MPa to 2527.22 MPa, while the torsional turns firstly increased from 15 to 20 turns and then decreased to 13 turns with the increase of strain. It can be observed in fig. 2 that with the increase of strain, a number of high density dislocation cells were observed in ferrite lamellae, which directly results in the increase of torsional strength. However, with the dislocation density increases above the yield point, as a result of crack formation that resulted in torsion delamination, the torsional turns decrease from 20 turns to 13 turns.

5. Conclusion

With the intention of investigating the microstructure evolution and mechanical properties of hypereutectoid steel wires in a large strain range, the high carbon steel wires with a carbon content of 1.02 wt.% is investigated. The conclusions are summarized as follows:

1. With the increase of the strain, the pearlite interlaminar spacing decreases gradually, and the ferrite and cementite coordinate deform violently. Additionally, the lamellar structure is bent or broken into pieces for those who perpendicular to the drawing direction, while it shows the fibrous morphology for those who parallel to the drawing direction.
2. A number of high density dislocation cells were observed in ferrite lamellae. It is worthy to be mentioned that, not only the ferrite lamellae, but also the cementite lamellae have a reduction in thickness with the increase of the strain. In addition, it has a different reduction level of thickness in different areas.
3. It can be seen that a number of nano-crystal cementite is distributed in cementite lamellae unevenly. Meanwhile, cementite lamellae were bent due to stress concentration caused by these dislocations. Different deformation degree of the cementite lamellae is associated to the amount of fractured cementite.
4. The strain rates go along with the increase of drawing turns, which results in the reduction of diameter of steel wire. With the strain increase from 0 to 2.81, torsional strength increases from 1126.83 MPa to 2527.22 MPa. The variation trend of torsional strength is increasing, which means the work hardening rate of hypereutectoid steel wires is improved with the increase of strain. The torsional turns firstly increased from 15 to 20 turns and then decreased to 13 turns with the increase of strain, which is related to the high density dislocation cells deformed after drawing.

Acknowledgement

This study was supported by the National Natural Science Foundation of China (51004039) and the Opening Project of Jiangsu Key Laboratory of Advanced Structural Materials and Application Technology (ASMA201602).

IOP Conf. Series: Materials Science and Engineering **436** (2018) 012018 doi:10.1088/1757-899X/436/1/012018

References

[1] Baraz V R, Sokolov A A and Belov A V 1995 *Met. Sci. & Heat. Treat.* **37(8)** 339-40
[2] Zelin M 2002 *Acta Mater.* **50(17)** 4431-47
[3] Tarui Toshimi, Maruyama Naoki and Eguchi Tatsuya 2002 *Struct. Eng. Int.* **12(6)** 33-40
[4] J. Embury and R. Fisher 1966 *Acta. Metall.* **14(2)** 147–59
[5] G. Langford and Metall 1970 *Mater. Trans. B* **1(2)** 465–77
[6] Langford G 1970 *Metall. Trans.* **1(2)** 465-477
[7] Porter D A, Easterling K E and Smith G D W 1978 *Acta. Mater.* **26(9)** 1405-22
[8] Lenasson C G and Bergstrom Y 1976 *The conference on the strength of metals and alloys* 141-45
[9] J. Languillaume, G. Kapelski and B. Baudelet 1997 *Acta. Mater.* **45** 1201–12
[10] Yu Ivanisenko, Lojkowski W and Valiev R Z 2003 *Acta. Mater.* **51(18)** 5555-70
[11] Lee M and Simkovich G 1988 *Metall. & Materials Trans. A* **19(8)** 2115-17
[12] Jae-Yong Chae, Jae-Hoon Jang and Guohong Zhang 2011 *Scr. Mater.* **65(4)** 245-48
[13] Read H G, Reynolds W T and Hono K 1997 *Scr. Mater.* **37(6)** 1221-30
[14] Q L Zhi, Ping J and Z H Wang 2008 *Mater. Lett.* **62(14):** 2825-27.

Microstructure and Bonding Strength of Aluminum Bronze on ASTM 1045 Steel by CMT welding

X M Zhang[1], K Sun[1], X H Gong[1], C M Jiang[1], J Zhou[1,2,*] and F Xue[1,2]

[1] Jiangsu Key Laboratory for Advanced Metallic Materials, Southeast University, Nanjing 211189, China
[2] Jiangsu Key Laboratory of Advanced Structural Materials and Application Technology, Nanjing Institute of Technology, Nanjing 211189, China

[*]jethro@seu.edu.cn

Abstract. The effect of welding parameters on metallographic microstructure of aluminum bronze, interfacial microstructure and bonding strength of aluminum bronze and steel substrate has been studied. A cladding of aluminum bronze was obtained on the surface of medium carbon steel by CMT welding and short circuiting transfer welding. The microstructure of aluminum bronze and the interfacial microstructure of the interface between aluminum bronze and steel substrate have been systematically investigated. Sample 1-4 are $\alpha + (\alpha + \gamma_2)$ microstructure of typical aluminum bronze. As for Sample 5, iron-rich particles and dendrites were distributed on the α-Cu crystal matrix. The bonding strength of aluminum bronze and steel substrate has been collected according to the destructive testing method of the bimetallic bonding strength of sliding bearing. Among the five samples, Sample 4 and Sample 5 had the highest bonding strength (528.17MPa and 511.39Mpa, respectively).The reason for the enhancement of the bonding strength was discussed. In the interface of aluminum bronze and steel substrate, the transition layer will get thicker when the interdiffusion distance between Fe and Cu elements is larger, it would be more conducive to increase the bonding strength.

1. Introduction

Copper alloys have been widely used in bearing materials due to their high fatigue strength and carrying capacity, excellent wear resistance, and good thermal conductivity, of which lead bronze and copper-lead alloy are the main copper alloy bearing materials[1-3]. Owing to the severe toxicity of Pb and its compounds, many countries have prohibited the use of lead-containing materials, so lead-containing copper-based sliding bearing materials have gradually been eliminated by the times. The developed high-strength lead-free bronze alloy has a higher allowable specific pressure than traditional lead bronze materials, and can well meet the requirements of high specific pressure and high outbreak pressure of modern engines[4-7].

Researchers have tried to produce aluminum bronze/steel bimetal bearing materials by powder metallurgy method, but have not succeeded. The problem of bond strength cannot be solved due to the inherent characteristics of aluminum bronze. Some countries (Italy, the United States) have focused their research on casting method, hoping to make breakthroughs. However, there have been no reports of large-scale industrialization applications[8].

Hence, promising welding methods are considered to produce aluminum bronze/steel bimetal bearing materials. In this paper, CMT (cold metal transfer) welding method was chosen to bond aluminum bronze to the steel substrate. CMT process is based on short circuiting transfer, or rather, on a deliberate, systematic discontinuing of arc. In a normal short circuiting transfer arc, the electrode is deformed while being dipped into the weld pool, and melts abruptly at high transfer arc current. In contrast to this, in CMT process, every time the short circuit occurs, the digital process control both interrupts the power supply and controls the retraction of the wire. The wire retraction motion assists droplet detachment during the short circuit. By ensuring minimal current metal transfer, the CMT process greatly reduces the amount of heat generated [9]. A cladding layer of aluminum bronze was obtained on the surface of steel substrate by CMT welding. The effect of different heat input on the microstructure and morphology of the interface was investigated. And the influence of the transition layer on the interfacial bonding strength was investigated too. A cladding layer of aluminum bronze was also obtained on the surface of steel substrate by short circuiting transfer welding as a control group.

2. Materials and method

The aluminum bronze wire used in this study was S215 (ERCuAl-A2). The diameter of wire was 1.2mm and the specific compositions were shown in Table 1. The ASTM 1045 steel was used as the steel substrate and its dimensions were 150 mm×150 mm×12 mm. Before welding test, the surface of substrate was polished to remove the oxidation layer. The welding wires were CMT welded and short circuiting transfer welded on the surface of steel substrate. The welding torch was perpendicular to the steel substrate. Multiple pass welding was conducted step-by-step. The schematic of welding test processes is illustrated in Figure 1. First, the welding torch moved about 120 mm along the length, then went back to the starting point immediately. After that, the welding torch walked about 13mm along the width and continued to weld along the length. A total of three welds were carried out to complete the welding process. Finally, these process were repeated in the same place once again to get a cladding layer with a height of about 4mm.

Table 1 Chemical composition of aluminum bronze alloy

Element	Content (wt%)
Al	8.67
Fe	1.04
Cu	90.28

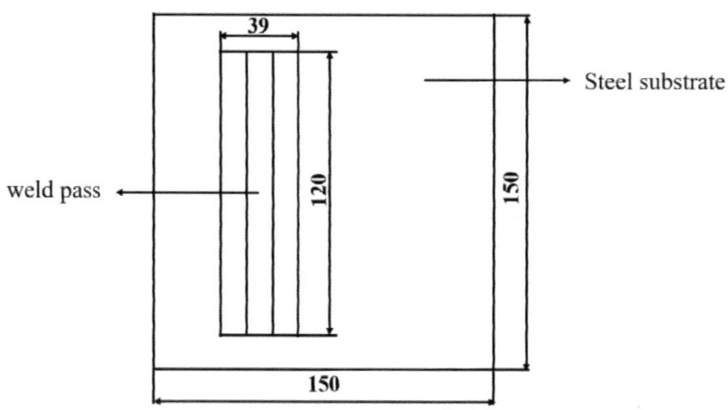

Figure 1 Schematic of welding test

The welding voltage is automatically matched with the welding current. The specific welding parameters were shown in Table 2. The scan speed was 30 cm·min^{-1} and the shielding gas was pure argon. The sample was cut into 10 mm×10 mm×16 mm with the thickness of aluminum bronze in the sample approximately 4 mm. Samples were polished beforehand and then corroded by ferric chloride aqueous hydrochloric acid. Finally, the microstructures of the aluminum bronze and interface were studied by the optical microscope (OM) and scanning electron microscopy (SEM). The chemical compositions were investigated by energy dispersive X-ray spectrometer (EDS).

Table 2 Welding parameters of aluminum bronze on steel substrate

Sample No.	Welding method	Current/A	Voltage/V
1	CMT	100	14.0
2		110	14.2
3		120	14.4
4		130	14.7
5	Short circuiting transfer	130	18.5

In this experiment, the bonding strength of aluminum bronze and ASTM1045 steel was tested according to the destructive testing method of the bimetallic bonding strength of sliding bearing which referred to the national standard GB/T12948-1991. The schematic and digital picture of sample is shown in Figure 2.

According to the regulations, the tensile speed was controlled by means of stress, and the applied value of stress was 10 N/(mm²·s). Until the specimen was ruptured, the maximum test load was recorded. Six replicate tests were performed on each group of samples. The relative error of the value was less than 10% and the bonding strength was calculated according to the following formula:

$$R=F_{max}/A$$

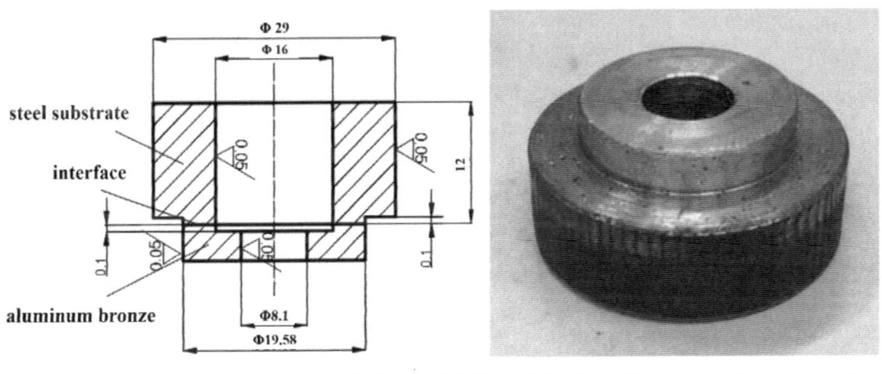

Figure 2 Schematic (left) and digital picture (right) of the bonding strength test specimen

3. Result

3.1 Microstructure results of aluminium bronze alloy

The metallographic microstructures of aluminum bronze alloy are exhibited in Figure 3. The microstructures of Sample 1-3 are the same as that of Sample 4. From Figure 3(a), all of Sample 1-4 are $\alpha + (\alpha + \gamma_2)$ microstructures of typical aluminum bronze, composed of black $(\alpha + \gamma_2)$ phase and white α-Cu matrix. Since the post-weld samples were all air-cooled, the β-phase was transformed into black $(\alpha + \gamma_2)$ phase by eutectoid decomposition process based on the Cu-Al binary phase diagram.

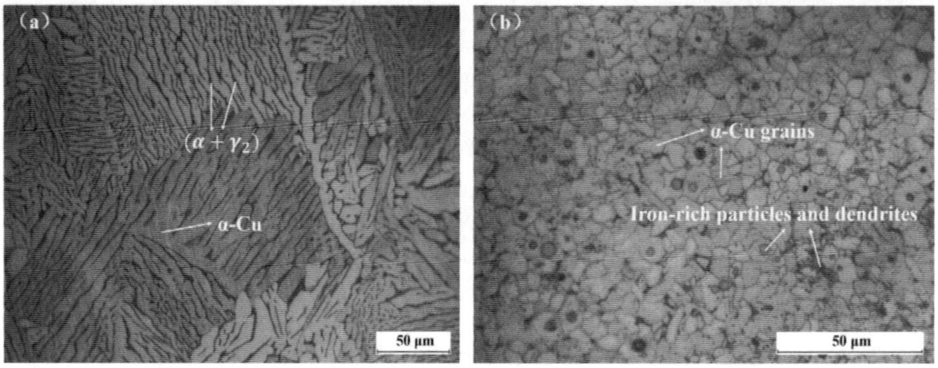

Figure 3 Microstructure of aluminum bronze: (a) Sample 4, (b) Sample 5

As can be seen from Figure 3(b), the microstructure of sample 5 is distinctly different from those of the other samples. Iron-rich particles and dendrites are distributed on the white α-Cu matrix. According to Table 3, it can be found that due to the dilution effect of Fe, the composition of Al was changed from original 8.67 wt% to present 7.16%, so the matrix was composed of single-phase α-Cu grains. Fe exceeded the solubility in copper-aluminum binary alloy and precipitated into iron-rich particles and dendrites.

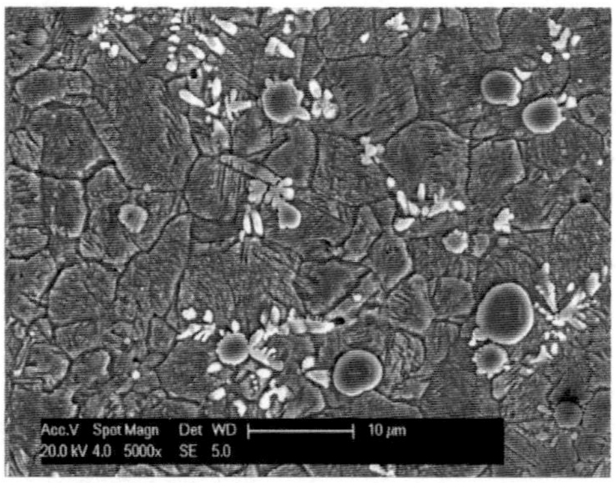

Figure 4 The SEM image of Sample 5

Table 3 The results of EDS at three positions in Figure 4 (wt%)

Position	Al	Fe	Cu
1	5.6	80.40	14.00
2	4.40	86.98	8.63
3	7.16	3.87	88.98

3.2 Interfacial structure of the aluminium bronze alloy/ASTM 1045 steel interface

The bonding strength between aluminum bronze and steel substrate depends on the microstructure at the interface. The interfacial microstructures of welding joints were firstly observed by OM, as shown in Figure 5. The interfacial structures of Sample 1-2 are the same as that of Sample 3. As shown in Figure 5(a), the demarcation line of the steel-copper of Sample 1-3 is clear and there is no obvious transition layer. As for Sample 4 and Sample 5, it can be found in Figure 5(b) and Figure 5(c) that dendrites grew perpendicularly to the interface. According to the results in Table 4, it is known that the dendrites are rich in iron.

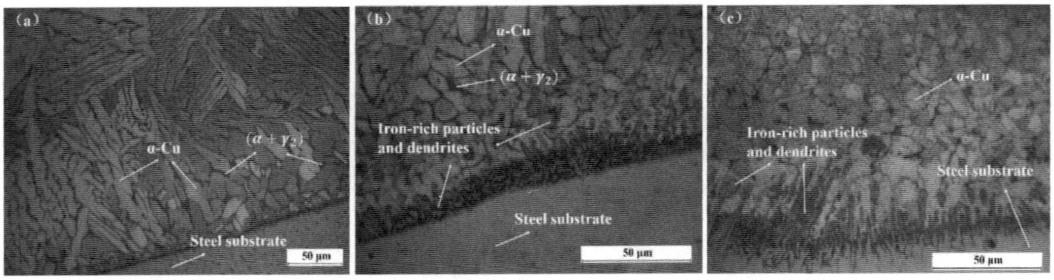

Figure 5 Interfacial structure at the interface: (a) Sample 3, (b) Sample 4, (c) Sample 5

The interfacial microstructure and composition of the interface were further analyzed by SEM and EDS. The SEM images of the interface of different experiment conditions and results of line scanning are shown in Figure 6. The results of EDS at different positions are displayed in Table 4.

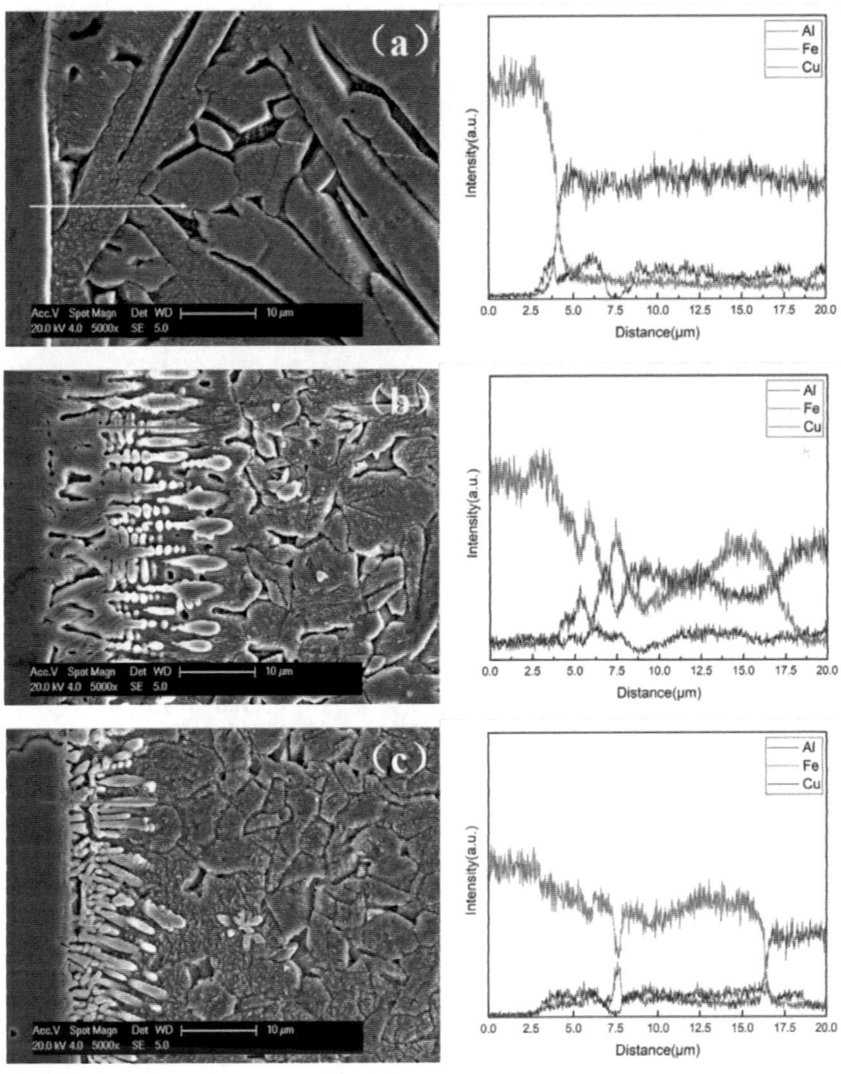

Figure 6 SEM and EDS analysis of the interfacial microstructure: (a) Sample 3, (b) Sample 4, (c) Sample 5

According to the results in Figure 6 and Table 4, there were no iron-rich dendrites at the interface of Sample 1-3. The Fe element just diffused into the copper alloy and the interaction range was only 2mm. On the interface of Sample 4 and Sample 5, the growth of iron-rich dendrites into the copper alloy was clearly observed. The interaction range was 15 mm and 13 mm, respectively.

Table 4 The results of EDS at different positions in Figure 6 (wt%)

Position	Al	Fe	Cu
1	7.85	5.05	87.10
2	8.48	4.13	87.39
3	4.76	82.22	13.01
4	9.00	72.99	18.01
5	8.23	76.26	15.51
6	7.61	6.19	86.21
7	7.12	78.85	1.34
8	10.59	72.96	16.45
9	8.28	4.04	87.64

3.3 The bonding strength of aluminium bronze alloy and ASTM 1045 steel

The average test results of bonding strength of different samples are shown in Figure 7. The bonding strength of Sample 1-3 were similar, being 430.63 MPa, 435.92 MPa, and 450.84 MPa, respectively. The bonding strength of Sample 4 and Sample 5 were 528.17 MPa and 511.39 MPa, which were much higher than those of Sample 1-3. It can be seen that under the same welding material, the different bonding strength was caused by different welding parameters.

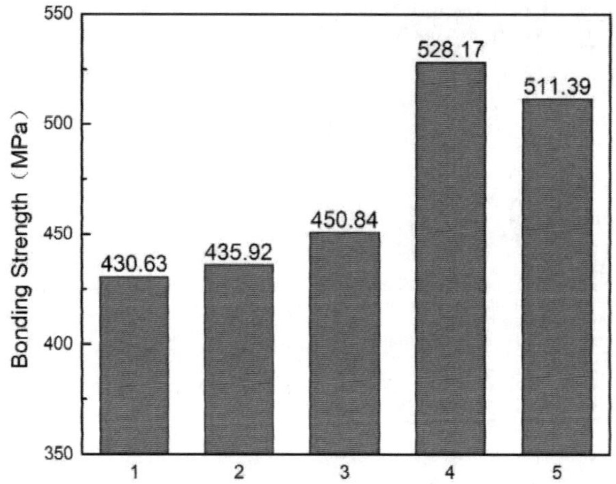

Figure 7 The bonding strength test results of aluminum bronze alloy and steel substrate

4. Discussion

4.1 Effect of different welding parameters on microstructure of aluminium bronze alloy

The metallographic microstructure and interfacial microstructure of Sample 1-3 were the same. This was because Sample 1-3 were all prepared by CMT welding, and the welding current was small, meaning that the heat input was relatively low. Therefore, only a small amount of iron diffused into the copper alloy and did not affect the metallographic microstructure of the copper alloy.

Although the interfacial microstructure of Sample 4 was similar to that of Sample 5, their metallographic microstructures were completely different. The welding current of Sample 4 and Sample 5 was both 130A. However, the heat input of Sample 5 prepared by short circuiting transfer welding was significantly higher than Sample 4 prepared by CMT welding. Due to the large amount of iron diffusing into the copper alloy, the composition of Al was changed from original 8.67 wt% to present 7.16%, so the matrix was composed of single-phase α-Cu crystal. Fe exceeded the solubility in copper-aluminum binary alloy and precipitated into iron-rich particles and dendrites. The heat input of Sample 4 was between Sample 3 and Sample 5, so there were dendrites growing perpendicularly to the interface. Meanwhile, the microstructure was still $\alpha + (\alpha + \gamma_2)$ microstructure of typical aluminum bronze.

4.2 Effect of different welding parameters on mechanical properties of aluminium bronze alloy
The bonding strength between aluminum bronze and steel substrate depends on the microstructure at the interface. The welding parameters were different, which led to various interfacial structures and caused the changes in bonding strength.

Due to the iron-rich dendrites, the interaction range at the interface of Sample 4 and Sample 5 was relatively large, 15 mm and 13 mm, respectively. There were no iron-rich dendrites at the interface of Sample 1-3 and the interaction range was only 2mm. Therefore, the bonding strength of Sample 4 and Sample 5 was much larger than that of Sample 1-3, and the bonding strength of Sample 4 is slightly larger than that of Sample 5. Samples 1-3 were very similar in terms of metallographic microstructure and interfacial structure, so bonding strength of them was also very close.

5. Conclusion

1) Sample 1-4 are $\alpha + (\alpha + \gamma_2)$ microstructures of typical aluminum bronze. As for Sample 5, iron-rich particles and dendrites were distributed on the α-Cu crystal matrix. The demarcation line of the steel-copper of Sample 1-3 was clear and there was no obvious transition layer. As for Sample 4 and Sample 5, dendrites grew perpendicularly to the interface.
2) A cladding layer of S215 aluminum bronze was obtained on the surface of the steel substrate by CMT welding and short circuiting transfer welding. The bonding strength of Sample 1-5 were 430.63MPa, 435.92MPa, 450.84MPa, 528.17MPa and 511.39Mpa, respectively.
3) In the interface of aluminum bronze and steel substrate, the transition layer will get thicker when the interdiffusion distance between Fe and Cu elements is larger, it would be more conducive to increase the bonding strength.

Acknowledgement
This study was supported by the National Natural Science Foundation of China (51004039) and the Opening Project of Jiangsu Key Laboratory of Advanced Structural Materials and Application Technology (ASMA201602).

References

[1] Davim J P 2000 *Journal of Materials Processing Tech* **100(1)** 273-7
[2] Kostornov A G and Fushchich O I 2005 *Powder Metallurgy & Metal Ceramics* **44(3-4)** 202-6
[3] Pratt G C 1968 *Metallurgical Reviews* **18(2)** 62-88
[4] Kestursatya M, Kim J K and Rohatgi P K 2003 *Materials Science & Engineering A* **339(1)** 150-8
[5] Heilmann P, Don J and Sun T C 1983 *Wear* **91(2)** 171-90
[6] Fukuda K 2003 *Journal of Japanese Society of Tribologists* **48(3)** 165-71
[7] Kamiya S and Tomikawa T 1999 *Japanese Journal of Tribology* **44(4)** 363-75
[8] Semnifer R J 2005 *Metals and Materials* **24(7)** 67-90
[9] Bruckner J 2005 *Welding Journal* **84(6)** 38-40

IOP Conf. Series: Materials Science and Engineering **436** (2018) 012020 doi:10.1088/1757-899X/436/1/012020

Defects Recognition in Selective Laser Melting with Acoustic Signals by SVM Based on Feature Reduction

D S Ye[1, 2, 3], Y H J Fuh[2], Y J Zhang[2], G S Hong[2] and K P Zhu[3 *]

[1]Department of Automation, University of Science and Technology of China, 230026, Hefei, China
[2]Department of Mechanical Engineering, National University of Singapore, 117575, Singapore
[3]Institute of Advanced Manufacturing Technology, Chinese Academy of Science, 213164, Changzhou, China

*Corresponding author: zhukp@iamt.ac.cn

Abstract. Defects among the selective laser melting(SLM) part hinder the development of the SLM process. This work provides an approach to conduct the monitoring and defect diagnosis by support vector machines (SVM) model using extracted features from acoustic signals. After training and testing with the linear SVM model, the result from the Fisher discriminant analysis (FDA) feature reduction performs optimal compared with those from the original features and the principal component analysis (PCA) feature reduction. The melted state monitoring and classification can be realized by simple discriminant model of SVM with extracted features after dimension reduction. The proposed method can be applied in the SLM process monitoring and defect diagnosis by acoustic signals with generalization.

1. Introduction

Defects occur within selective laser melting(SLM) additive manufacture parts, which interfere with the product applications among aerospace, health care and industry. The defects are mainly from the underheating and overheating melting process over overhang structures or unsteadily melting sections. Process monitoring of the SLM process is the premise for process feedback control to solve defects [1]. Therefore, it is necessary to implement the SLM process monitoring technique for improving the SLM part quality.

SLM process is accompanied with optical, acoustic, electron, and other weak signals. Many researches have been conducted to monitor the SLM process. Kruth's group developed a series of coaxial monitoring and control approaches with combined cameras and photodiodes. The melt pool size and thermal dynamics were analysed to match the SLM process [2, 3]. Lane's group executed the thermography measurement process with multiple off-axial optical sensors. Temperature and brightness of the melt pool as well as the spatter observation were captured for the process monitoring [4]. Rieder et al. demonstrated an online ultrasonic monitoring system to investigate the dynamic layer building process. The residual stresses and possible porosity were indicated by evaluating the backwall signals and ultrasonic velocities [5]. Ye et al. collected the acoustic signal for monitoring the characteristics by a microphone. There is a good mapping between audible signals and laser power as well laser scanning speed during the SLM process [6]. Melted states have been recognized by deep learning algorithm [7].

All the signals from image or acoustic information can represent the status of the melting. Monitoring with coaxial optical setups easily affects the captured signals by lens characteristics, while monitoring with off-axial optical setups is influenced by the sensor installation site of clamping position, angle, and distance. The acoustic signal can predict the penetration state during processing since there is a relationship between acoustic signal features and penetration states [8]. As a flexible setup and non-destructive method, acoustic methods offer a wide variety of detectability and provide information on defect location and shape [9]. Acoustic signals have been successfully applied in the monitoring of the SLM process using deep belief networks, while more accurate and simpler algorithms can be used to detect the defects.

In terms of acoustic signals during the processing, the modes are time-varying signals due to the changing of spatters on the powder, temperature around the melt pool, and formations around the melting site. Acoustics signals are non-stationary signals. Both low frequency and high frequency bands have information from acoustic signals. Wavelet package transform (WPT) can decompose acoustic signals into approximations and details. Acoustic signal features in each frequency range indicate the relationship between failure modes and components [10]. Components of the decomposed acoustic signals are analyzed by parameter statistics. To make less computation and robust representation for different melted states, feature selection for feature dimension reduction is necessary. Principal component analysis (PCA) and Fisher linear discriminant analysis (FDA) are used to make the feature dimension reduction. Comparisons methods with original features, PCA feature selection and FDA feature selection are carried out to evaluate the present approach. In the final, conclusions are provided for this work.

2. Feature extraction

2.1 Wavelet packet decomposition of acoustic signals

Acoustic signals from the SLM process are time varying and non-stationary. With the non-stationary signal from the melting conditions, WPT can decompose signals into different frequency bands with different resolutions. Features of acoustic signals in different depths are sensitive to different melted states and insensitive to the stationary working conditions. The acoustic signal measurement is $x(t)$ for the WPT analysis and parameter statistics. With different decomposed depths of the WPT, the measurement is [11]

$$Wx_n(l,k) = \langle x(t), \mu_n(t,l,k) \rangle = \int_{-\infty}^{+\infty} x(t)\mu_n(t,l,k)dt \quad (1)$$

where μ_n is the n_{th} wavelet packet function with $n = 0,1,2...2^l$. l is the level and k is the location parameter. $\langle \ , \ \rangle$ donates the inner product. The melting condition is able to be reflected with different wavelet packet decomposed signals $Wx_n(l,k)$.

2.2 Feature parameter statistics

Melting condition can be reflected by the acoustic signal information included in different wavelet packet nodes. Statistical parameters are useful indicators for extracting the melting condition. In this work, absolute mean amplitude (F_1), variance value (F_2), skewness factor(F_3), kurtosis(F_4), square root amplitude(F_5), crest factor(F_6), impulse indicator(F_7), shape factor(F_8), and cepstrum factors(F_9 and F_{10}) for signals in each frequency range are calculated respectively. In the non-stationary signal processing, ten parameter statistical methods with physical meaning are applied [12-14].

$$F_1(l,n) = \frac{1}{K}\sum_{k=1}^{K}\left|Wxn(l,k)\right| , F_2(l,n) = \frac{1}{K}\sum_{k=1}^{K}(Wx_n(l,k))^2, F_3(l,n) = \frac{1}{K}\sum_{k=1}^{K}(Wx_n(l,k))^3, F_4(l,n) = \frac{1}{K}\sum_{k=1}^{K}(Wx_n(l,k))^4 \quad (2)$$

$$F_5(l,n)=\left(\frac{1}{K}\sum_{k=1}^{K}Wx_n(l,k)^2\right)^{1/2},\quad F_6(l,n)=\frac{\max|Wx_n(l,k)|}{(\frac{1}{K}\sum_{k=1}^{K}(Wx_n(l,k))^2)^{1/2}},\quad F_7(l,n)=\frac{\max|Wx_n(l,k)|}{\left(\frac{1}{K}\sum_{k=1}^{K}(|Wx_n(l,k)|)^{1/2}\right)^2}$$

$$F_8(l,n)=\frac{\left(\frac{1}{K}\sum_{k=1}^{K}(Wx_n(l,k))^2\right)^{1/2}}{\frac{1}{K}\sum_{k=1}^{K}|Wx_n(l,k)|},\quad F_9(l,n)=\frac{1}{K}\sum_{\omega}\log|\Phi_l^n(\omega)|e^{2\pi\omega i/K},\quad F_{10}(l,n)=\frac{1}{K}\sum_{\omega}\log|\Phi_l^n(\omega)|e^{4\pi\omega i/K}$$

where $\Phi_l^n(\omega)$ is Fourier transformations of wavelet packet decomposing signals $Wx_n(l,k)$. With the WPT decomposition and parameter statistics, the melting measurement $x(t)$ can be extracted as

$$F_i=(F_1,F_2,F_3,F_4,F_5,F_6,F_7,F_8,F_9,F_{10}),\ i=1,2,...,10 \qquad (3)$$

3. Feature selection and classification

3.1 PCA for feature selection

After wavelet packet decomposition and feature statistical analysis, the total number of features extracted by acoustic signals is 160 dimensions. To decrease the computation complexity and increase the robust classification ability, the extracted feature dimension needs to be reduced for realizing the SLM process monitoring. PCA method uses the covariance of the input vector to project the features from high-dimensional space into the low-dimensional space. It finds the projection method that best represents the original data in the minimum mean square sense.

The normalized feature is $\mathring{F}_i=F_i-\overline{F_i}$,where $\overline{F_i}$ is the mean values of feature F_i. Sample covariance matrix C is $\mathring{F}^T\mathring{F}/T$, where T is the feature number of measured acoustic signals. By PCA, $\mathring{F}$ was transferred into $S=Z^T\mathring{F}$, where Z is $N\times N$ orthogonal matrix and z_j is the j_{th} eigenvector of the sample covariance matrix C. The eigenvalue λ_j of the covariance matrix C is $Cz_j=\lambda_j z_j$, $j=1,...,N$. PCA ratio η represents that the principal component dominants the whole information $\eta=\left(\sum_{j=1}^{M_\eta}\lambda_j\Big/\sum_{j=1}^{N}\lambda_j\right)^{1/2}$.

3.2 FDA for feature selection

The PCA is to find the direction to represent the principal information effectively. FDA method is carried out to find the direction of effective classification. It distinguishes various classes with maximizing the between-class distance and minimizing the within-class distance. The mean vectors of the p_{th} features of different samples are calculated by $m_i^q=n_q^{-1}\sum_{K\in\omega_q}F_i^K$, $q=1,2,...$, where ω_q belongs to class q. The summary of the within-class covariance matrix of class ω_q is $S_{\omega_q}=\sum_{K\in\omega_q}(F_i^K-m_i^q)(F_i^K-m_i^q)^T$. The between-class covariance matrix is $S_B=\max(m_i^a-m_i^b)^2,\ a,b\in q$. For making the two classes as far as possible and the same class as clustered as possible, the ratio of between-class distance and within-class distance is defined $J_i=S_B/S_{\omega_d}$. Rank J_i from the largest to the smallest and select the top M_ε between-class distance and within-class distance ratios. The FDA ratio ε is taken with J_i as $\varepsilon=\sum_{i=1}^{M_\varepsilon}J_i\Big/\sum_{i=1}^{N}J_i$

3.3 Classification with Support Vector Machines (SVM)

The SVM algorithm searches for an optimal hyperplane that separates the data into classes. The discriminant function for the optimal hyperplane is

$$f(x) = sign(\sum_i^D \gamma_i y_i G(x_i, x) + b) \quad (4)$$

where y_i is the observation corresponding to the sample x_i. D is the sample number and γ_i is the Lagrange multipliers. $G(x_i, x)$ is the kernel functions and b is a bias term. The function is solved by

$$\max_{\lambda}(\sum_{i=1}^D \gamma_i - \sum_{i=1}^D \sum_{j=1}^D \gamma_j y_i y_j G(x_i, x))\gamma_i, \quad \sum_{i=1}^D \gamma_i y_i = 0, \quad 0 \le \gamma_i \le E \quad (5)$$

where E is the normalization value for separating classes.

4. Experiment setup

Experiments are conducted with an own designed SLM system as shown in Figure 1(a). The SLM system is composed by a laser, a powder spreading facility, and a controller panel. The acoustic monitoring system for experiments includes a 3780C1 PCB microphone and a SIRUSm data acquisition system. The frequency response of the acoustic signal acquisition system is $0 \sim 200\ KHz$ and is mounted $2cm$ above the platform (Figure 1(b)). Single tracks are conducted on $50\mu m$ -thickness 304L stainless steel powder. Parameters are changed to obtain the states of overheating, normal, and underheating.

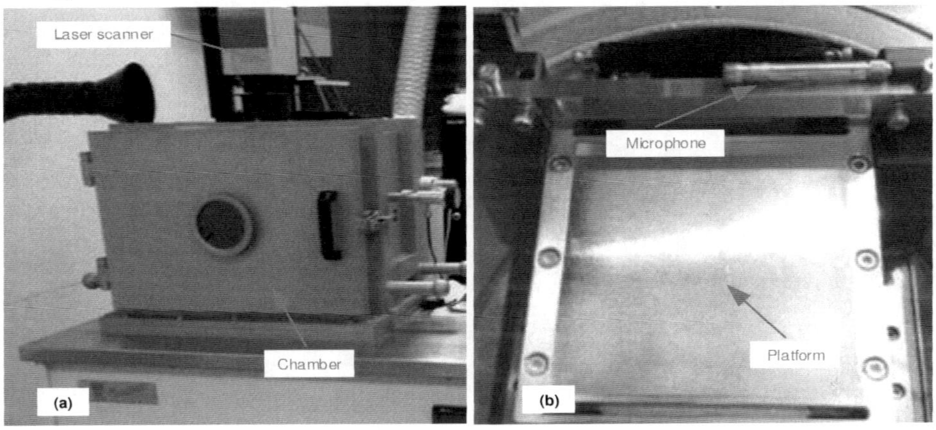

Figure 1. (a) SLM machine for the melting process, and (b) experiment setups with an acoustic signal acquisition system.

All the single tracks are observed with an Olympus microscope with 200 times enlargement. There are three typically features of the single tracks as shown in Figure 2. Balls take shape due to underheating caused by lack of energy input as illustrated in Figure 2(a). In Figures 2(b) and 2(c), satellites on the single tracks come from the powder and liquid spatter separately. The wettability is good from the energy density and satellites are mostly from the powder spatter that has no effect on the adjacent remelting. Melted tracks as Figure 2(b) are normal melted states.

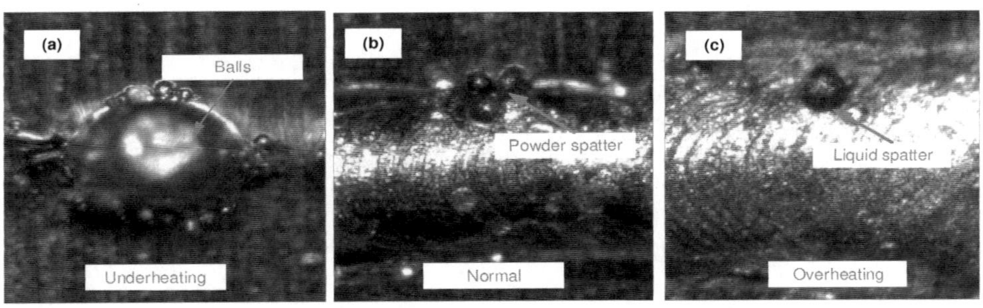

Figure 2. Melted states of single tracks from different melting conditions.

Satellites larger than $50\mu m$ cannot be sufficiently melted and tend to hinder the paving roller movement or affect the following layer. Single tracks with larger than $50\mu m$ satellites are defined as medium overheating states. The melted tracks with energy density larger than that for medium overheating ones are overheating states. The melted states between normal and underheating states are medium underheating ones. Single tracks are conducted from different energy input to obtain underheating, medium underheating, normal, medium overheating, and overheating states separately.

Figure 3 shows the acoustic signals and the corresponding appearance of the quality of one melted track from the SLM process. The single-track image has been multiplied by 50 times. In this work, $10mm$ single tracks are conducted with each melted state for $7\ s$. Test experiments are conducted with different scanning speed and laser power for $2.3\ s$ as listed in Table 1.

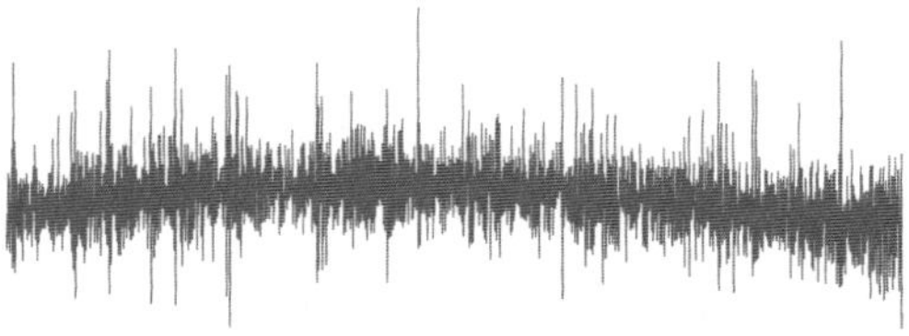

(a)

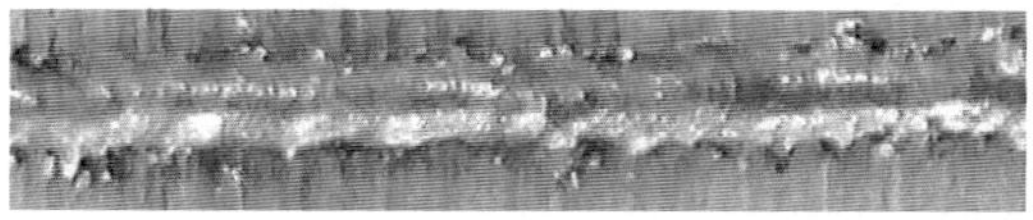

(b)

Figure 3. Acoustic signal detected in the PBF-L process.

Table 1. Parameters for the test experiments.

Test	Scanning speed (mm/s)	Laser power (W)	Melted state
1	100	50	Underheating
2	700	100	Medium underheating
3	110	100	Normal
4	100	150	Medium overheating
5	40	100	Overheating

5. Experiment results and discussions

In order to recognize the melted states, the sample window is set with $10k$ signal points. 140 train samples and 46 test samples are obtained for each melted state. Acoustic signals are decomposed with 4-level WPT of Daubeches 3 wavelet and signals reconstructed from the fourth-level wavelet package coefficient are chosen as the features for parameter statistics (Figure 4). Statistical parameters have 160 dimensions after feature extraction. PCA and FDA are performed for the feature dimension reduction.

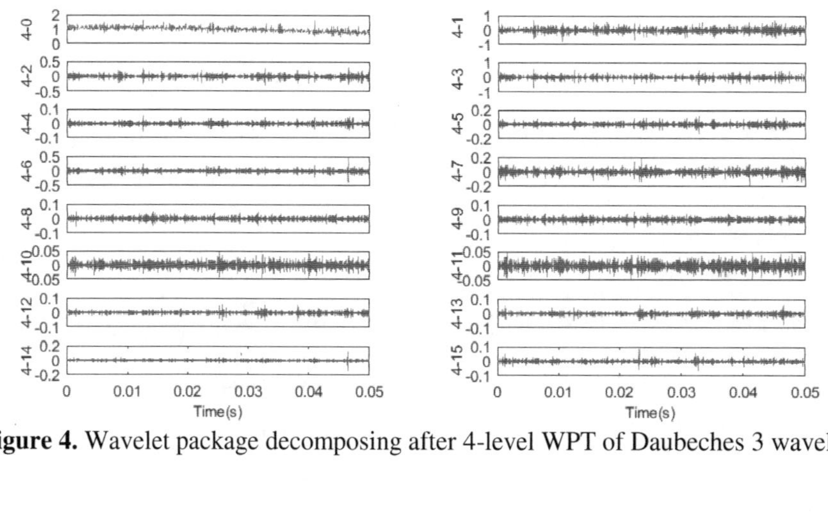

Figure 4. Wavelet package decomposing after 4-level WPT of Daubeches 3 wavelet.

34 features are selected after FDA dimension reduction with FDA ratio $\varepsilon = 0.99$ and 20 features are left after PCA dimension reduction with PCA ratio $\eta = 0.99$. Selected features are trained and tested by SVM with linear kernel function. Results of five-melted-state classification are shown as Table 2.

With empirical analysis, melted state defects diagnosis is simple and accurate with appropriate extracted features. PCA feature dimension reduction method works as a feature extraction method instead of real feature selection. Some physical nature of the acoustic signal features is lost during the PCA feature extraction. Therefore, recognition with PCA feature selection performs weak result of 55.35% classification rate in the melted state classification. With the original features, feature dimensions are larger than training samples. Although the discrimination rate is 89.13%, overfitting easily emerges and there is less generalization in the SLM process monitoring. FDA as linear discriminant method performs 77.99% rate for the classification. FDA method suits the recognition of melted state since the monitoring model has generalization in the state classification.

Table 2. Test results with different dimension features.

Dimension reduction method	Test result
Original features	89.13%
PCA	55.35%
FDA	77.99%

6. Conclusion

This work proposes an innovative method to realize the SLM process monitoring. Acoustic signals are decomposed into different frequency ranges by WPT. Signals at each frequency range are calculated with ten kinds of parameter statistical methods and extracted features are reduced by PCA and FDA. By the linear SVM classification model, the optimal result is from statistical parameters after FDA feature dimension reduction. The proposed approach with FDA feature dimension works well in the discrimination of melted states. It can be carried in the monitoring and defect diagnosis of the SLM process.

Acknowledgement

The authors would like to acknowledge the supports from the China Scholarship Council, Chinese Academy of Sciences 100 Talents Program, and National University of Singapore through research attachment for this work. This work is also supported by the National Additive Manufacturing Innovation Cluster, Singapore through a PEP project funding.

References

[1] S.K. Everton, M. Hirsch, P. Stravroulakis, R.K. Leach and A.T. Clare, Mater. Design, **95**: p. 431- 445, (2016).
[2] J.P. Kruth, G. Levy, F. Klocke and T.H.C. Childs, CIRP. Ann. Manuf. Technol, **56**, (2007).
[3] S. Clijsters, T. Craeghs, S. Buls, K. Kempen and J.P. Kruth, Int. J. Adv. Manuf. Tech, **75**, 5: p.1089-1101, (2014).
[4] B. Lane, E. Whitenton and S. Moylan. *Proc SPIE.*(2016).
[5] H. Rieder, A. Dillhöfer, M. Spies, J. Bamberg and T. Hess. *AIP Proc.*(2015).
[6] D. Ye, Y. Zhang, K. Zhu, G.S. Hong and J.Y.H. Fuh, *Adv Energy Sci Eng II*, p. 1315-1320, (2017).
[7] D. Ye, G.S. Hong, Y. Zhang, K. Zhu, and J.Y.H. Fuh, Int. J. Adv. Manuf .Tech: p. 1-11, (2018).
[8] S. Song, H. Chen, T. Lin, D. Wu and S. Chen, J. Mater. Process. Tech, **234**: p. 33-44, (2016).
[9] S. M. Bree,. no. March (2015).
[10] R. Khamedi, A. Fallahi and A. Refahi Oskouei, Mater. Design, **31**, 6: p. 2752-2759, (2010).
[11] R. Yan, R.X. Gao and X. Chen, Signal. Process, **96**: p. 1-15, (2014).
[12] K. Zhu, G.S. Hong and Y.S. Wong, Mach. Sci. Technol, **12**, 3: p. 348-369, (2008).
[13] D. You, X. Gao and S. Katayama, IEEE. T. Ind. Electron, **62**, 1: p. 628-636, (2015).
[14] C. Li, R.V. Sanchez, G. Zurita, M. Cerrada, D. Cabrera and R.E. Vásquez, Mech. Syst .Signal. Pr, **76-77**: p. 283-293, (2016).

Electrochemical Assembling of Functionalized PEDOT Thin Films with Excellent Electroactivity and Superhydrophobicity

Zhi Geng[1], Zhengwei You[2,*] and Bo Zhu[3,*]

[1,2]State Key Lab for Modification of Chemical Fibers and Polymer Materials& College of Materials Science and Engineering, Donghua University, 2999 North Renmin Road, Songjiang, Shanghai, 201600, China
[3]College of Materials Science and Engineering, Shanghai University, 99 Shangda Road, BaoShan, Shanghai, 200444, China

[*]E-mail: zyou@dhu.edu.cn; bozhu@shu.edu.cn

Abstract. We synthesized dodecyloxymethyl functionalized EDOT monomers and fabricated homogeneous thin films via cycle voltammetry electrochemical deposition method on platinum and ITO substrates. The PEDOT-C12 thin films showed good electrochemical activity. Meanwhile, by increasing the cyclic voltage-scan number during electrochemical deposition process, the surface roughness was also enhanced. The large surface roughness and occurrence of micro-nano surface morphologies would lead to higher water contact angles for the PEDOT-C12 thin films. The water contact angle for the PEDOT-C12 thin films with 10 deposition cyclic voltage-scans was 160° and showed superhydrophobicity. The achieved superhydrophobic PEDOT-C12 thin films would have potential applications in corrosion protecting for metal electrodes in aqueous environment.

1. Introduction

Conducting polymers, which possess similar conductivity to metal, have gained great development since the first discovery in 1977.[1-4] Among them, poly(3,4-ethylenedioxythiophene) (PEDOT) materials have been commonly used in the fields of chemical and biological sensing, energy storage, bioelectronics due to the low oxidant voltage and good aqueous stability.[5-9] However, further applications would not only demand excellent electric conductive, but also extra functional properties based on various requirements. Introducing proper functional groups into PEDOT chain would expand the potential application fields for PEDOT materials. For instance, carboxylic acid functionalized EDOT thin films would specific bind with target cells via antibody-antigen interaction;[10] phosphorylcholine functionalized PEDOT thin films displays high resistance to nonspecific enzyme/cell binding.[11]

Besides, superhydrophobic thin films have gained great interests from researchers, which possess water contact angle higher than 150° and sliding angle smaller than 10°.[12-15] Superhydrophobic thin films show excellent repellence to water and would prevent it from penetrate into the materials. Combination superhydrophobicity with conducting polymer thin films would be effective method for fabricating electroactive corrosion protecting thin films based on current researches.[16, 17] Two main factors influence the hydrophobicity of conducting polymer thin films. First is the surface energy of the materials. Low surface energy is necessary for fabricating superhydrophobic thin films. Meanwhile, proper nano morphologies on the surface would also determine the hydrophobicity of thin films, in

IOP Conf. Series: Materials Science and Engineering **436** (2018) 012021 doi:10.1088/1757-899X/436/1/012021

which micro-nano surface morphologies have been proved to be the key point for superhydrophobic surface. Therefore, synthesizing functional EDOT monomer with low surface energy and realizing controllable nano assembly of the functionalized PEDOT are necessary for fabricating superhydrophobic PEDOT thin films.

Herein, we synthesized EDOT monomer with dodecyloxymethyl groups into the EDOT chain (PEDOT-C12). Cyclic voltammetry Electrochemical deposition method was used to realize the homogeneous assembling of PEDOT-C12 both on platinum and ITO conductive substrates. The achieved PEDOT-C12 thin films showed good charge exchange property based on cyclic voltammetry measurement. Meanwhile, the PEDOT-C12 thin films showed high surface roughness by increasing the deposition scans. The high surface roughness would lead to enhancing of water contact angles, and the PEDOT-C12 thin films with 10 deposition scans showed superhydrophobicity. The achieved PEDOT-C12 thin films possessed both electroactivity and superhydrophobicity, which would have potential application in corrosion protecting of electrodes in aqueous environment.

2. Results and discussion

To achieve the superhydrophobic PEDOT thin films, it is necessary to synthesize functional EDOT monomer with low surface energy. Starting from the hydroxyl modified EDOT (EDOT-OH), dodecyloxymethyl functionalized EDOT (EDOT-C12) could be synthesized by a simple chemistry pathway. The long alkyl chain would provide relatively low surface energy for the achieved EDOT-C12 monomer.

Electrochemical deposition was carried out to polymerize the EDOT-C12 monomer. Acetonitrile with 100mM Tetrabutylammonium Hexafluorophosphate (TBAPF6) was chosen to be the electrochemical solution. 5mM EDOT-C12 monomer was dissolved into the electrochemical solution. Platinum and ITO conductive substrates were used as working electrode and platinum net was used as counter electrode. Typical cycle voltammetry electrochemical deposition method was chosen by applying cyclic potentials from -0.6-1.45 V (vs Ag/Ag+). Figure 1 illustrate the deposition curve for the PEDOT-C12 thin films on ITO and platinum substrates. Obviously, the current become larger with high deposition scans, indicating more EDOT-C12 monomers were deposited both on ITO and platinum substrates.

The achieved PEDOT-C12 possessed excellent electrochemical properties. Cycle voltammetry measurement was used to testify the electrochemical properties for the PEDOT-C12 thin films in acetonitrile solution with 100mM TBAPF6. Figure 2 showed the CV curve for the PEDOT-C12 thin films with various scanning rate. PEDOT-C12 thin films showed good electrochemical properties on ITO and platinum substrates. Meanwhile, the oxidant current in the cycle voltammetry measurement was increased linearly with higher scan rate, indicating that the cycle voltammetry measurement was activated on the PEDOT-C12 thin films, rather than in the electrolyte.

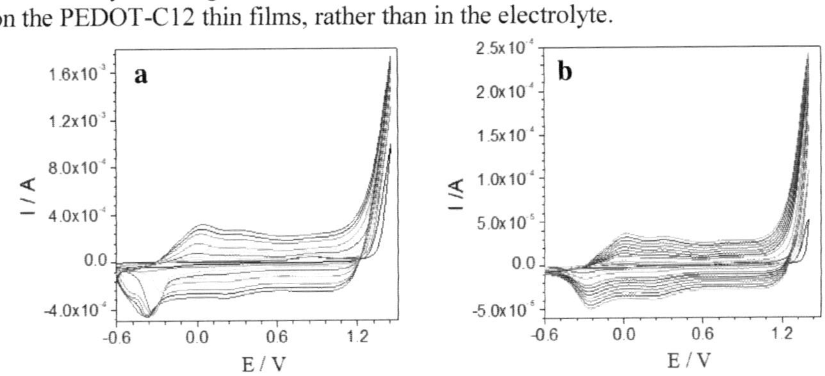

Figure 1 Cycle voltammetry electrochemical deposition curves for EDOT-C12 on ITO (a) and platinum (b) conductive substrates. The conductive of the thin films increased with high scanning numbers.

IOP Conf. Series: Materials Science and Engineering **436** (2018) 012021 doi:10.1088/1757-899X/436/1/012021

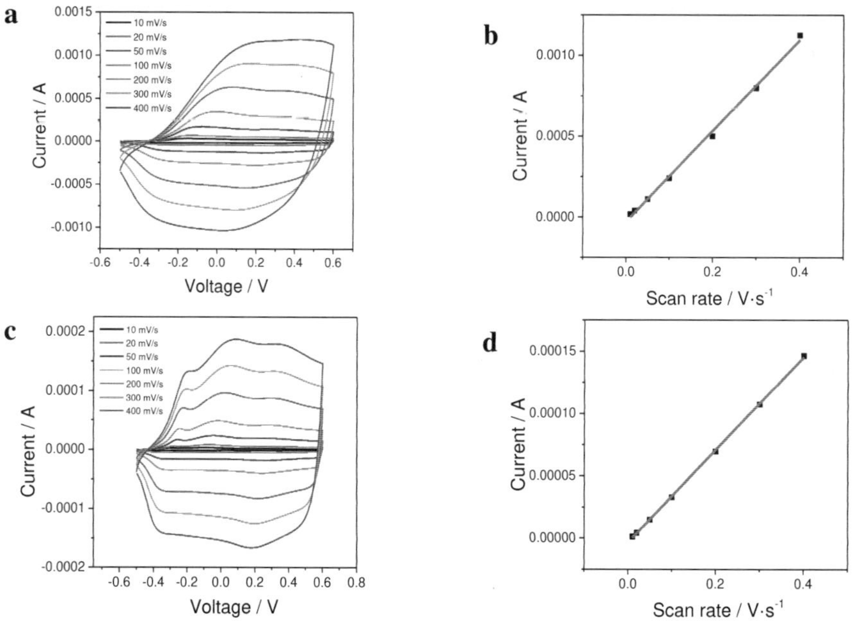

Figure 2. Cycle voltammetry measurement curves for PEDOT-C12 thin films on ITO (a) and platinum (c) conductive substrates. The scanning rate was 10, 20, 50, 100, 200, 300 and 400mV, respectively. The relationship between current and scan rate for the PEDOT-C12 thin films on ITO (b) and platinum (d) conductive substrates in the cycle voltammetry measurement

Besides of the electrochemical activity, the increasing of deposition scan number would also affect the surface roughness of the PEDOT-C12 thin films due to the nano assembling of monomers during the electrochemical deposition process. Atomic force microscopy (AFM) was used to measure the surface morphologies and surface roughness for the PEDOT-C12 thin films. As shown in Figure 3, the surface was quite smooth for the PEDOT-C12 thin films with 1 and 2 deposition cyclic voltage-scans from -0.6 to 1.45 V. The surface roughness was 6.7 and 15.1 nm, respectively. Keep increasing the scan number, micro-nano surface morphologies were occurred on the PEDOT-C12 thin films. The surface roughness was increased to 226.3 and 497.7 nm for PEDOT-C12 thin films with 5 and 10 scan numbers.

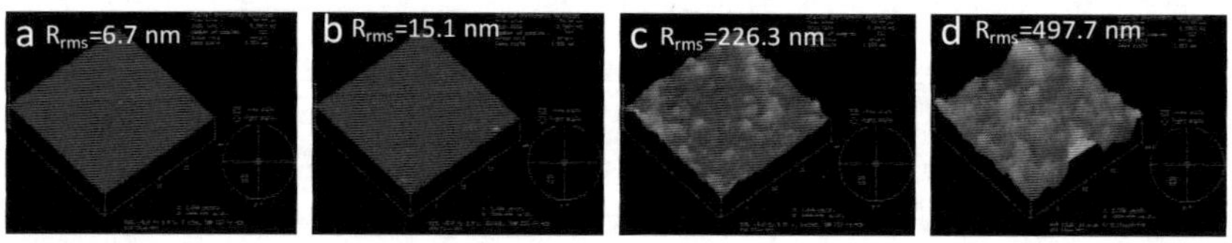

Figure 3. Roughness (as calculated on AFM measurements) for EDOT-C12 deposited on ITO plates within a) 1, b) 2, c) 5 and d) 10 deposition cyclic voltage-scans from -0.6 to 1.45 V.

The increasing of surface roughness and the occurrence of micro-nano surface morphologies would contribute to the superhydrophobicity, as the micro-nano surface morphologies could preserve small air chamber and suit for Cassie-Baxter model. Figure 4 showed the wettability of the achieved PEDOT-C12 thin films. Obviously, with the increasing of surface roughness, the water contact angles were also enhanced. The PEDOT-C12 thin films with 10 deposition cyclic voltage-scans possess the largest surface roughness and the water contact angle of the thin film was 160°, which showed excellent superhydrophobicity.

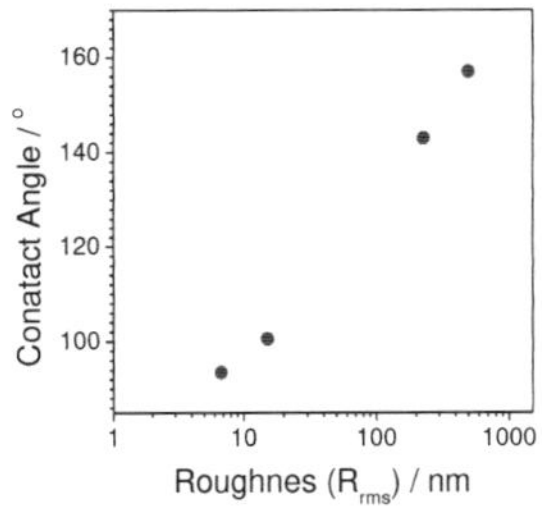

Figure 4. Water contact angles of PEDOT-C12 thin films and its dependence on surface roughness.

3. Conclusion

In this work, we designed dodecyloxymethyl functionalized EDOT monomers and fabricated homogeneous thin films via cycle voltammetry electrochemical deposition method on platinum and ITO substrates. The PEDOT-C12 thin films showed good electrochemical activity. Meanwhile, by increasing the cyclic voltage-scan number during electrochemical deposition process, the surface roughness was also enhanced. The large surface roughness and occurrence of micro-nano surface morphologies would lead to higher water contact angles for the PEDOT-C12 thin films. The water contact angle for the PEDOT-C12 thin films with 10 deposition cyclic voltage-scans was 160° and showed superhydrophobicity. The achieved PEDOT-C12 thin films possessed both superhydrophobicity and excellent electrochemical activity, which would have potential applications in corrosion protecting for metal electrodes in aqueous environment.

4. Experimental Section

4.1. Materials

Hydroxyl functionalized 3,4-dimethoxythiophene (99%+), NaH, 18-crown-6 ether and bromoethane were purchased from Alfa Aesar (China). Tetrabutylammonium Hexafluorophosphate (99%) and LiClO4 were purchased from Energy Chemical Reagent Co., Ltd (Shanghai, China). Sodium dodecyl sulfate (99%) was purchased from J&K Chemical Reagent Co., Ltd (Shanghai, China). Acetonitrile (99%), dichloromethane (99%) THF(99.5%+) were purchased from Ling Feng Chemical Reagent Co., Ltd (Shanghai, China) and used without further purification. ITO glass and platinum were purchased from Shenzhen Huanan Technology Company.

4.2. Fabricating of EDOT-C12 monomer

A mixture of NaH (3.48 g, 145 mmol) and 18-crown-6 ether (0.767 mg, 2.9 mmol) dissolved in anhydrous THF (70 mL) was stirred under nitrogen at 0 °C. 2,3-Dihydrothieno[3,4-b][1,4]dioxin-2-yl)methanol (EDOT-OH) (5 g, 29 mmol) in 30 mL of anhydrous THF was added. The mixture was stirred at room temperature for 1 h and then cooled down to 0 °C. Bromoethane (3.792 g, 34.8 mmol) was added droplet within10 min. After stirring at room temperature overnight, the mixture was diluted with 200 mL NH_4Cl aqueous solution. The mixture was extracted with dichloromethane (200 mL) three times. The combined organic phases were washed with water, dried over $MgSO_4$, and evaporated in vacuo. This yielded a colorless oil (8.19 g, 83%). 1 H NMR (500 MHz, CDCl 3 , δ): 6.33-6.31 (m, 2H), 4.32-4.28 (m, 1H), 4.25 (dd, J = 11.5, 2.0 Hz, 1H), 4.06 (dd, J= 11.5, 7.5 Hz, 1H), 3.68 (dd, J = 10.0, 4.5 Hz, 1H), 3.59 (dd, J = 10.0, 6.0 Hz, 1H), 3.49 (t, J = 7.0 Hz, 2H), 1.60-1.55 (m, 2H), 1.35-1.22 (m, 18H), 0.88 (t, J = 7.0 Hz, 3H). 13 C NMR (500 MHz, CDCl 3 , δ): 141.6, 141.5, 99.7, 99.5, 72.6, 72.1, 69.1, 66.3, 31.9, 29.7, 29.6, 29.6, 29.6, 29.5, 29.4, 29.4, 26.0, 22.7, 14.1;

4.3. Electrochemical polymerization

Electrochemical polymerizaion of PEDOT-C12 thin films performed using Autolab PGSTAT128N potentiostat (Metrohm Autolab) and a three-electrode electrochemical cell, with a platinum net as the counter electrode, ITO glass and platinum conductive substrates as working electrode. Electrochemical solvents were prepared by dissolving EDOT-C12 (5mM) and Tetrabutylammonium Hexafluorophosphate (0.1 M) in acetonitrile. EDOT-OH (10mM) were dissolved in water containing 50mM SDS and 100mM $LiClO_4$. PEDOT-OH thin films were pre-coated on ITO glass by applying cycle voltammetry from -0.6 V to 1.13 V vs Ag/AgCl standard electrode. PEDOT-C12 thin films were prepared on pre-coated ITO glass and platinum conductive substrates by applying cycle voltammetry from -0.6 V to 1.45 V vs Ag/Ag+ standard electrode. Each measurement was calibrated using a standard ferrocene/ferrocenium redox system.

4.4. Characteristic

The Cycle voltammetry test were measured by Autolab PGSTAT128N potentiostat (Metrohm Autolab) and a three-electrode electrochemical cell, in which Pt would act as counter electrode, PEDOT-C6 thin films as working electrode and Ag/Ag+ electrode as reference electrode. Acetonitrile with 100mMTetrabutylammonium Hexafluorophosphate was used as electrolyte.

Surface morphologies of the thin films were measured by Atomic force microscope (Agilent5500).

Water contact angles was measured by contact angle measurement system (Theta Lite 101, Attention).

Acknowledgements

B.Z. acknowledges financial support from the NSFC (21474014) and the NSF of Shanghai (14ZR1400200). Z.G. thanks financial support from the NSFC (21704013) and China Postdoctoral Science Foundation (2017M611416).

References

[1] C. K. Chiang, C. R. Fincher, Y. W. Park, A. J. Heeger, H. Shirakawa, E. J. Louis, S. C. Gau and A. G. MacDiarmid, 1977 *Physical Review Letters* **39** 1098.
[2] F. Zhao, Y. Shi, L. Pan and G. Yu, 2017 *Accounts of Chemical Research* **50** 1734.
[3] Y.-Z. Long, M.-M. Li, C. Gu, M. Wan, J.-L. Duvail, Z. Liu and Z. Fan, 2011 *Prog Polym Sci* **36** 1415.
[4] C. Li, H. Bai and G. Shi, 2009 *Chemical Society Reviews* **38** 2397.
[5] D. Khodagholy, J. Rivnay, M. Sessolo, M. Gurfinkel, P. Leleux, L. H. Jimison, E. Stavrinidou, T. Herve, S. Sanaur, R. M. Owens and G. G. Malliaras, 2013 *Nat Commun* **4** 2133.
[6] J. Rivnay, S. Inal, B. A. Collins, M. Sessolo, E. Stavrinidou, X. Strakosas, C. Tassone, D. M. Delongchamp and G. G. Malliaras, 2016 *Nat. Commun* **7** 11287.
[7] A. Williamson, J. Rivnay, L. Kergoat, A. Jonsson, S. Inal, I. Uguz, M. Ferro, A. Ivanov, T. A. Sjostrom, D. T. Simon, M. Berggren, G. G. Malliaras and C. Bernard, 2015 *Adv. Mater.* **27** 3138.
[8] J. Francis Fennell, H. Hamaguchi, B. Yoon and T. M. Swager, 2017 *Sensor* **17** 982.
[9] R. Emanuelsson, M. Sterby, M. Strømme and M. Sjödin, *J. Am. Chem. Soc.* **139** 4828.
[10] J. Sekine, S. C. Luo, S. Wang, B. Zhu, H. R. Tseng and H. h. Yu, 2011 *Adv. Mater.***23** 4788.
[11] B. Zhu, S.-C. Luo, H. Zhao, H.-A. Lin, J. Sekine, A. Nakao, C. Chen, Y. Yamashita and H.-h. Yu, 2014 *Nat. Commun.* **5** 4523.
[12] Z. Geng and J. He, 2015 *Adv. Mater. Inter.* **2** 1500196.
[13] Z. Geng, J. He, L. Xu and L. Yao, 2013 *J. Mater. Chem. A* **1** 8721.
[14] Z. Geng and J. He, 2014 *J. Mater. Chem. A* **2** 16601.
[15] A. B. D. Cassie and S. Baxter, 1944 *Trans. Faraday Soc.* **40** 546.
[16] H. Xu, J. Liu, Y. Chen, J. Tang and Z. Zhao, 2016 *J. Appl. Polym. Sci.* **133** 44248.
[17] K. C. Chang, H. I. Lu, C. W. Peng, M. C. Lai, S. C. Hsu, M. H. Hsu, Y. K. Tsai, C. H. Chang, W. I. Hung, Y. Wei and J. M. Yeh, 2013 *ACS Appl. Mater. Inter.* **5** 1460.

IOP Conf. Series: Materials Science and Engineering **436** (2018) 012022 doi:10.1088/1757-899X/436/1/012022

Study on the Equivalent Pressure Coefficient of the Viscosity Model Shaped by Ultra-thin Injection Molding

Yin Sufeng[1,a], Fang Lin[1,b] and Huang Hai[1,c]

[1]Department of Mechanical and Electrical Engineering, City College of Dongguan University of Technology, Dongguan, 523419, China

[a]1042784725@qq.com; [b]494812652@qq.com; [c]447620419@qq.com

Abstract. In ultra-thin injection molding, with the increase of pressure, the nonlinear relationship between pressure and viscosity becomes more and more obvious. The pressure dependence of Cross-WLF viscosity model can no longer reflect the actual production.So,this paper based on the free volume theory, expounds how the pressure affects the melt viscosity in the Cross-WLF viscosity model, and based on the variations of the specific volume in the PVT melt curve, demonstrates that the pressure dependence of melt viscosity is significantly nonlinear in the ultra-thin high-pressure injection molding, which is inconsistent with the traditional linearity. In order to reflect the actual production better, this paper defines the equivalent pressure coefficient of the viscosity model and justifies it in the production of the ultra-thin light guide plate.

1. Introduction

The Polymer Rheology defines that the viscosity of high polymer melt increases with the rise of pressure, otherwise it decreases in normal cases. In regular injection molding, the injection pressure is generally below 100Mpa, thus, most researches tend to focus on the pressure dependence of melt viscosity under low pressure. However, in ultra-thin injection molding where the injection pressure is up to 300Mpa, the injection is speedy and the product precision is demanding, the pressure dependence of melt viscosity will have a significant impact on the molding process. Given its uniqueness, relevant literature is rarely seen.

Westover[1] discovers that the rise of pressure enables the viscosity of polystyrene to increase by more than 100 times provided that the shearing rate is constant. Son Y[2] find that the viscosity of polystyrene is increased by 134 times on the condition that the temperature stays at 196℃ and the pressure boosts from 0 to 126Mpa while the viscosity of polystyrene is only increased by 14 times when the temperature stands at 149℃ and the pressure changes by the same degree. In light of the relations between the fractional free volume and the temperature, Kadijk S E [3]comprehensively explains the pressure dependence of the viscosity by using zero shear viscosity and shear thinning practice in Arrhenius-WLF Equation and Carreau Equation respectively. Fatma Dinc[4] employs free volume theory and develops a new viscosity model. According to this model, zero shear viscosity, constant shear stress viscosity, and constant shear rate viscosity are all in a linear relationship with the fractional free volume, in which the viscosity declines as the fractional free volume increases. P. Van Puyvelde[5] controls the capillary in the fashion of back pressure and improves the pressure's impact on the viscosity by way of correcting the pressure at the inlet. Gu Zhengwei[6] argues that since zero shear viscosity η_0 is obtained in low pressure (below 50Mpa) environment, the pressure's impact on the viscosity must be included if the pressure of injection molding is high. As a result, he corrects the reference point of zero shear viscosity in the Cross Model and verifies how important the pressure dependence of melt viscosity is to the simulation analysis of thin-wall items.

In numerous literatures have been discussed the pressure's impact on the viscosity within the low-pressure scope. As for high-speed ultra thin molding technology, it is commonly recognized that the pressure dependence of the viscosity is crucial to the molding process. But subject to the limitation of laboratory conditions, few literatures make reference.

2. pressure dependence of viscosity in Cross-WLF seven-parameter model

The Cross-WLF seven-parameter model[7] is

$$\eta(\dot{\gamma},T,P)=\frac{\eta_0(T,P)}{1+(\eta_0\dot{\gamma}/\tau^*)^{1-n}} \tag{1}$$

where

$$\eta_0(T,P)=D_1\exp(\frac{-A_1(T-T^*)}{A_2+(T-T^*)}),\quad T^*=D_2+D_3\cdot P,\quad A_2=\tilde{A}_2+D_3\cdot P$$

n is a non-Newtonian index; τ*, a material constant; D_1, the shear stress prior to the shear thinning of melt; D_2, the glass transition temperature under low pressure; D_3, a coefficient affecting the viscosity and pressure; T*, the temperature at which the melt transforms into glass; A1, $\tilde{A}_2$ factors related to temperature

The pressure's impact on viscosity in the model is determined by the product of $D_3\times P$. The recommended values of several plastic D_3 are shown in table 1. The value of D_3 is small. When the pressure of injection molding is low, the product of $D_3\times P$ would also be small and have unnoticeable impact on viscosity. When the pressure of injection molding is high, the product of $D_3\times P$ would be large. Therefore, in the ultra thin high-pressure injection molding, the value of D_3 is significant to the molding process.

Table 1 D_3 values of several common materials

Materials	D_3(K/Pa)
PC	1.9×10^{-7}
PMMA	1.7×10^{-7}
PS	2.3×10^{-7}

3. Free Volume Theory and Pressure Coefficient of the Cross-WLF Viscosity Model

Fox and Flory[8] believe that the total volume (V) of liquid or solid has two components: one is (V_0) the volume occupied by molecules; the other is (V_f) the free volume, that is,

$$V=V_0+V_f \tag{2}$$

The free volume is scattered in the matter in the form of "cavity". The high polymer molecule chain adjusts its conformation through the rotation and displacement of the free volume. As the segment is frozen and the free volume is also frozen and kept at a fixed value, the free volume of high polymer in glassy state starts defrosting and swelling when the temperature rises up to the transition point of glassy state, as shown in Figure 1.

When T<Tg, the total volume of high polymer is[9]-[10]:

$$V_g=V_f+V_0+\left(\frac{dV}{dT}\right)_g T_g \tag{3}$$

V_f-free volume under glassy state

V_0-occupied volume of glassy high polymer under absolute zero

Similarly, when T<Tg, the volume of high polymer is:

$$V_r = V_g + \left(\frac{dV}{dT}\right)_r (T - T_g) \tag{4}$$

The free volume under the rubbery state when the temperature is T

$$(V_f)_T = V_f + (T - T_g)\left[\left(\frac{dV}{dT}\right)_r - \left(\frac{dV}{dT}\right)_g\right] \tag{5}$$

Where the difference between the rubbery state and glassy state $\left(\frac{dV}{dT}\right)_r - \left(\frac{dV}{dT}\right)_g$ is the expansion rate of free volume above T_g the expansion coefficients of high polymer above or below Tg are respectively as follows:

$$a_r = \frac{1}{V_g}\left(\frac{dV}{dT}\right)_r, a_g = \frac{1}{V_g}\left(\frac{dV}{dT}\right)_g \tag{6}$$

The expansion coefficient of free volume adjacent to Tg is the difference of expansion efficient of high polymer above and below T_g,

$$a_f = a_r - a_g \tag{7}$$

The fractional free volume at a temperature above the glass transition temperature is expressed:

$$f_T = f_g + a_f (T - T_g) \quad (T \geq T_g) \tag{8}$$

Where f_g is the factional free volume of glassy high polymer.

According to free volume theory, the very reason the viscosity of melt decreases is that the temperature rises above the glass transition temperature and the fractional free volume expands.[11]

The pressure dependence of high polymer viscosity can be summarized as follows: under pressure, the free volume within macromolecule materials dwindles and the molecule chain becomes less active, making the glass transition temperature rise up. Suppose the high polymer has State 1 （P_1, T） and State 2 （P2, T）, let isothermal compression be the equal of increasing the glass transition temperature of polymer, here comes:

$$\alpha_f (T_{g2} - T_{g1}) - K_f (P_2 - P_1) \tag{9}$$

K_f-the isothermal compression coefficient of free volume
as a result

$$\alpha_f (T_2 - T_1) = K_f (P_2 - P_1) \tag{10}$$

$$\frac{dT_g}{dP} = \frac{K_f}{\alpha_f} = D_3 \tag{11}$$

The equation 11 indicates that in the Cross-WLF viscosity model, D_3 is regarded as a constant because K_f is thought to be irrelevant with P. But actually it is not true. For example, as shown in the Figure 2, Panlite L-1225L(PC) which is an exhibition of the relationship between the specific volume and the pressure and temperature, there is a non-linearity between the pressure and the specific volume when the temperature is constant and this non-linearity intensifies as the pressure stays at high level in particular. Thus, when the melt is in compression, K_f is not a constant value. As the pressure rises, the value of K_f is on the decline, so is the value of D_3.

IOP Conf. Series: Materials Science and Engineering **436** (2018) 012022 doi:10.1088/1757-899X/436/1/012022

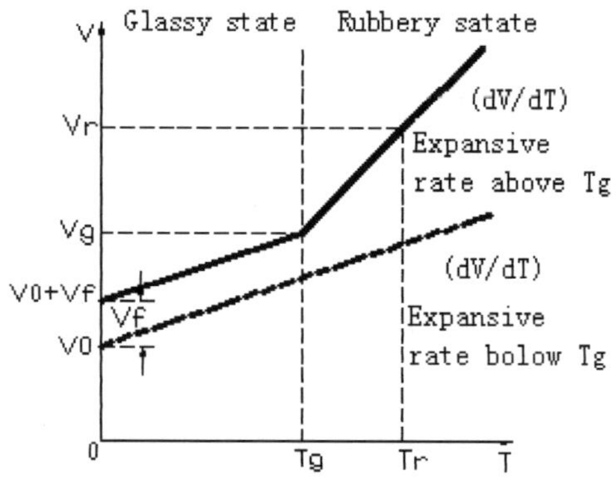

Figure.1 Melt volume expansibility with temperature changes

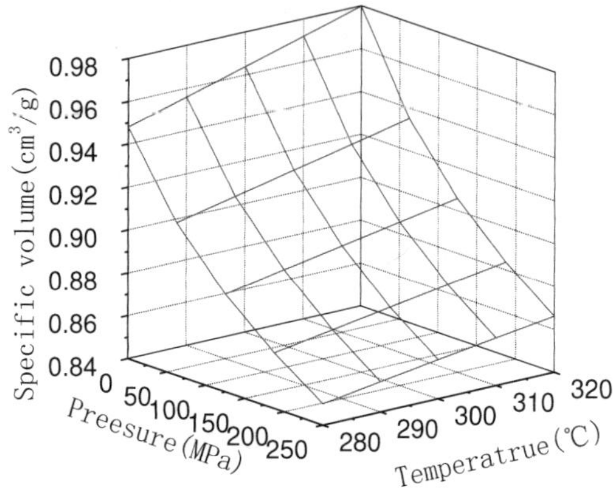

Figure.2 Panlite L-1225L (PC) PVT curve

4. D_3, the Equivalent Pressure Coefficient of the Cross-WLF Viscosity Model

In order to linearly represent the melt compressibility so that the linear representation of the pressure's impact on viscosity is realized in the Cross-WLF viscosity model, the equivalent compression gradient (shown in dotted line) of melt volume is selected in this paper to linearly calculate the amount of melt compression. Its value selection principle is to make the area of S_1 and S_2 in the figure equal. The objective is to compensate the positive and negative deviation of the compressed volume under high pressure and low pressure and appropriately reflect the actual compression amount of melt. The relationship between the compression amount of specific volume of Panlite L-1225L (PC) and the pressure is shown in Figure 3.

To master the compression law of free volume, macromolecules and free volume in melt are seen as two different series elastomers of elasticity modulus. These two elasticity modulus are all the function of the pressure P, that is $k_f(P)$ and $k_r(P)$, as shown in Figure 4.

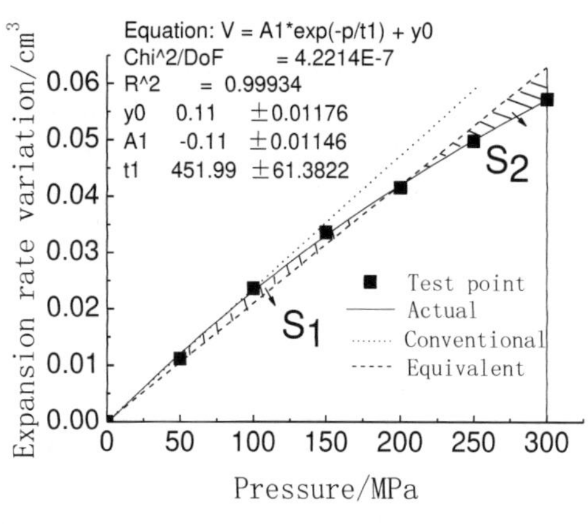

Figure.3 The relationship between the compression amount of specific volume of Panlite L-1225L (PC) and the pressure

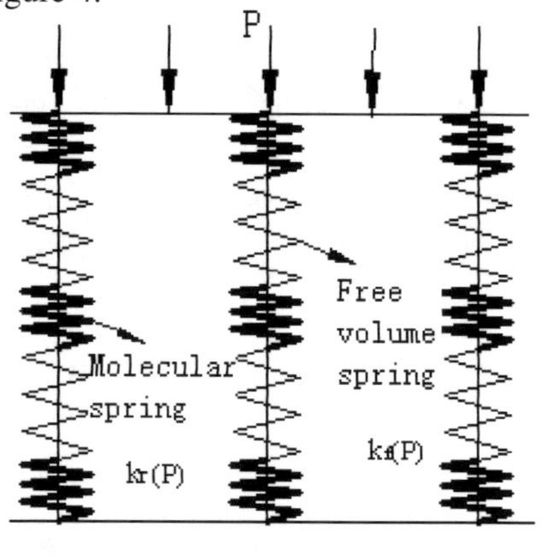

Figure.4 Polymer molecules and free volume compression model

Suppose the elasticity modulus of two elastomers are in direct proportion when the melt is under pressure, namely,

$$k_f(P) \propto k_r(P) \text{ and } k_f(P) = \xi k_r(P) \tag{12}$$

then comes

$$k_f(P) + \xi k_r(P) = \bar{k}(P) \tag{13}$$

$$k_f(P)\left(1+\frac{1}{\xi}\right)=\bar{k}(P) \tag{14}$$

$$k_f(P)=\left(\frac{\xi}{1+\xi}\right)\bar{k}(P) \tag{15}$$

Based on the study on the free volume, macromolecule expansibility and compressibility in literature, ξ is valued at 1.68[12].

With the experimental data curve of PC PVT, based on equation 6

$$a_f=\Delta a=a_r-a_g \tag{16}$$

the free volume expansibility of PC is:

$$a_f=6.2\times10^{-4}-1.9\times10^{-4}=4.3\times10^{-4} \tag{17}$$

with equation 10, the equivalent pressure impact coefficient is:

$$\overline{D}_3=\frac{k_f(P)}{\alpha_f}=\frac{\left(\frac{\xi}{1+\xi}\right)\bar{k}(P)}{\alpha_f} \tag{18}$$

With equation 18, the equivalent compression coefficient of the viscosity of Panlite L-1225L is shown in Figure 5.

The parameters of Cross-WLF viscosity model of the material Panlite L-1225L (from moldflow2012) are ahown in table 2. The comparison of viscosity curves is shown in Figure 6 on the conditions that the pressure-affecting coefficient D_3 takes the recommended value 1.9e-007 and equivalent value $\overline{D}_3$. The pressure in the Figure 6 is valued at 50 Mpa, 150 Mpa, and 300 Mpa. The Figure 6 shows that the impact the equivalent pressure-affecting coefficient D_3 exerts on the viscosity is less than the recommended value 1.9e-007. Therefore, it is more appropriate to use equivalent pressure-affecting coefficient to make calculations and actual production in the ultra-thin high pressure injection molding.

Table 2 parameters of Cross-WLF viscosity model of Panlite L-1225L

parameters	n	τ *	D_1	D_2	D_3	A_1	$\tilde{A}_2$
values	0.17	7.8e005	2.9e010	417	1.9e-007	25.9	61.2

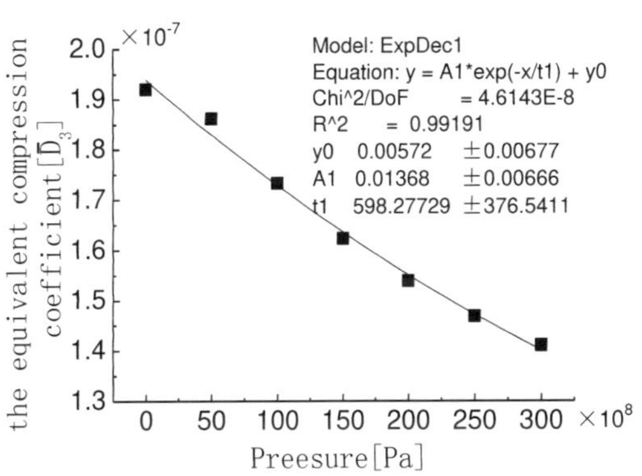

Figure.5 the equivalent compression coefficient pressure of the viscosity of Panlite L-1225L

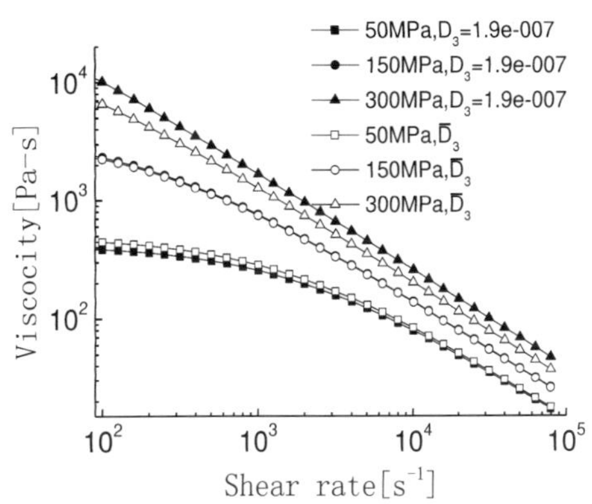

Figure.6 The different effects of equivalent coefficients and recommended values

5. Verification of the Equivalent Pressure Coefficient

Injection molding machine is TOYO SI-80IV. Test process parameters are shown in table 3.There is a sample of 0.4mm and 3in ultra-thin light guide plate for mobile phone manufactured by Dongguan Yidong Electronics Co., Ltd for Motorola company. Its dimension is shown in Figure 7. The injection molding machine is TOYO SI-V series full-automatic injection molding machine. The materials of test-piece are GF8002（PC）and the mold for test is shown in Figure 8.

Table 3 Test process parameters

materials	Melt temperature	mold temperature	injection speed	screw diameter	maintained pressure	Time of maintaining	Mold-locking force
Panlite L-1225L	310℃	110℃	500mm·s^{-1}	20 mm	60%	0.5 s	70 Tom

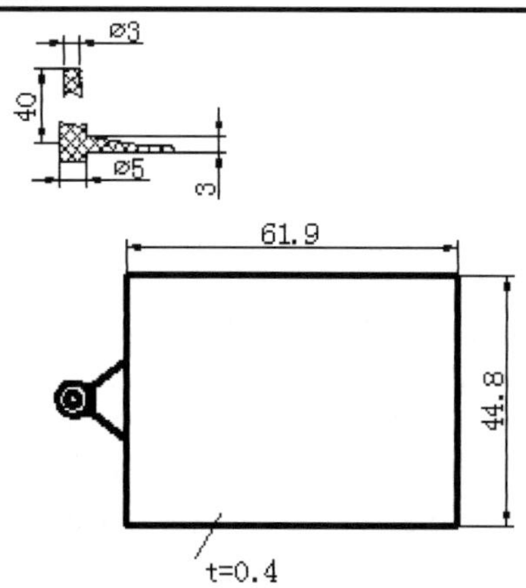

Figure.7 Geometric dimensions of samples

Figure.8 The mold for test

As shown in Fig.9(a), short injections are conducted on test piece as per the process parameters in Table 3 under the conditions that the injection pressure is set at 150Mpa, 200Mpa, 250Mpa, and 300Mpa respectively. Then as shown in Fig. 9(b), simulation is made in accordance with process parameters when the equivalent pressure-affecting coefficient D_3 is respectively set at 1.623e-007,1.540-007,1.496e-007, and 1.410e-007 selected based on the injection pressure. Finally, the results of simulation are shown in Fig.9 (c) when the pressure-affecting coefficient D_3 is set 1.9e-007 and the process parameters are kept unchanged.

The compassion between the short injection results and simulation results in Fig. 9 shows that the results of filled area of cavity simulated through using equivalent pressure-affecting coefficient D_3 are extremely close to the results of short injection, particularly, under the high-pressure injection state.

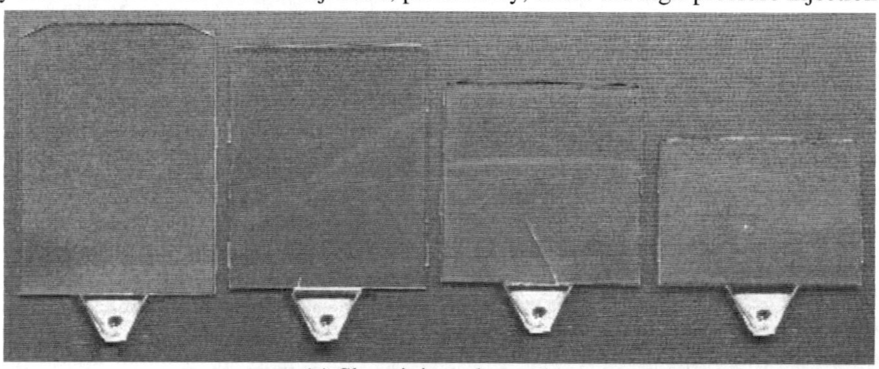

(a) Short injected samples

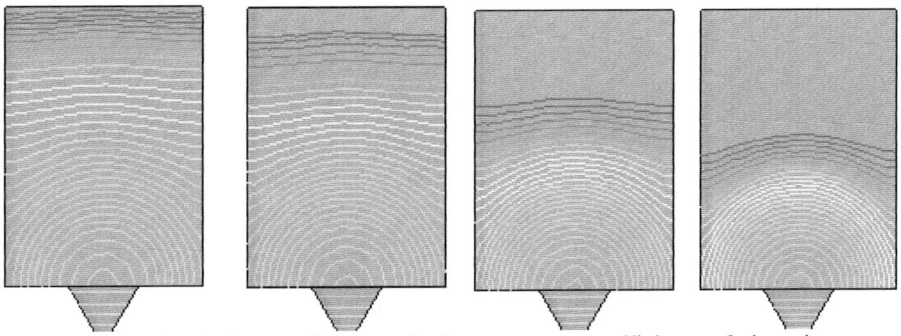

(b)The simulation results of equivalent pressure coefficients of viscosity

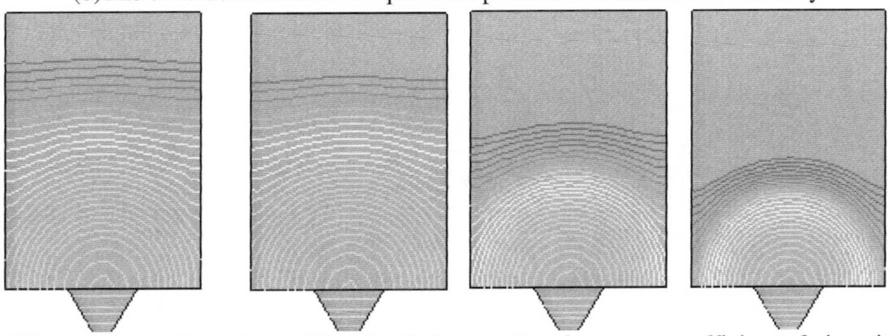

(c)The recommended values of the simulation results of pressure coefficient of viscosity

Figure. 9 The short injecting verification experiments of the equivalent pressure coefficients of viscosity

The above two experiments have demonstrated that the pressure dependence of viscosity model in ultra-thin injection molding must be considered because the injection pressure is forceful. It has been spotted from the practice that the pressure-affecting coefficient of Cross-WLF model in ultra-thin injection molding is largely deviated from the audit results under high pressure molding state. In light of the equivalent pressure-affecting coefficient proposed in this paper, the shortcomings of the Cross-WLF model in ultra-thin simulation have been satisfactorily addressed.

6. Summary

This paper concludes that the expansion ratio and compression ratio of free volume in melt are not proportional. The law of pressure's impact on viscosity cannot be scientifically reflected by equalizing the pressure's impact on viscosity with the linear increase of glassy transition temperature of the melt. To compensate the shortcomings of viscosity model in ultra-thin injection molding, on the basis of the free volume theory, the melt compression under high or low pressure, the compression ratio of appropriate and constant specific volume is selected under specific pressure to accurately reflect the non-linear changes of the specific volume as the pressure changes. On such basis, the concept of equivalent pressure-affecting coefficient is developed to rationally reflect objective reality. Through the comparison between value analysis and practice, the simulated results of equivalent pressure-affecting coefficient fit relatively well to the production practice.

References

[1] Westover R F 1996 Effect of hydrostatic pressure on polyethylene melt rheology *POLYMER ENGINEERING & SCIENCE* **1(1)** 14-20
[2] Son Y 2009 Measurement of pressure dependence on the shear viscosity of polymer melts *Journal of Polymer Research* **16(6)** 667-71.
[3] Kadijk S E and Van den Brule BHAA 1994 On the pressure dependency of the viscosity of

molten polymers *Polym Eng Sci* **34 (20)** 1535-46
[4] Fatma Dinc Tomas Sedlacek and Cumali Tav 2014 On the non-newtonian viscous behavior of polymer melts in terms of temperature and pressure-dependent hole fraction *Journal of Applied Polymer Science*, **131(12)** 986-89
[5] Van Puyvelde P Vananroye A and Hanot A S 2013 On the Pressure Dependency of the Bagley Correction *International Polymer Processing* **28 (5)** 558-62
[6] Gu Zhengwei, Yang Shenhua and Kou Shu qing 2002 Study on the Viscosity Models of the Flowing Melt Plastics in the Injection Mould *Die & Mould Industry* **261(11)** 13-15
[7] Wu Qiye and Wu Jing-an 2002 *Rheology of Polymer Materials*（Peking : Higher Education Press）p121-25
[8] Fox T G and Flory P J 1950 *J.Appl.Phys* **21(6)** 581-84
[9] Van Puyvelde P, Vananroye A and Hanot A S 2013 On the Pressure Dependency of the Bagley Correction *International Polymer Processing* **28(5)** 558-62
[10] Halász L and Huszár A 2011 Free Volume from Pressure and Temperature Dependent Viscosity and from PVT Measurements for Homo- and Copolymers *International Polymer Processing* **26(4)** 403-06
[11] Zhang Zhongliang, Yan Xiaoli and Gao Shu 2012 Influence of Pressure and Temperature on the Free Volume of Polystyrene Studied by Positron *J. Wuhan Univ. (Nat. Sci. Ed.)* **58(5)** 377-80
[12] Dlubek G, Shaikh M Q and Krause Rehberg R 2007 A effect of free volume and temperature on the structural relaxation in polymethylphenylsiloxane: Apositron lifetime and pressure-volume-temperature study *THE JOURNAL OF CHEMICAL PHYSICS* **126(2)** 092202

Study on the Influence of Process Parameters on Interface Strength of Film-insert Molding

Zhangyong Hu[1,2], Annan Zhang[1], Zhiyong Wang[1] and Yang Li[1,*]

[1]State Key Laboratory of Material Processing and Die & Mold Technology, Huazhong University of Science and Technology, Wuhan, 430074, P. R. China
[2]Huanggang Normal University, College of Mechanical and Electrical Engineering, Huanggang City, HuangGang, 438000, P. R. China

*liyang@hust.edu.cn

Abstract. By the controlling of variable, the influences on the film interface strength that film-line molding process had were investigated. The interfacial bonding strength under the circumstance where various of processing parameters (temperature, film thickness, injecting speed) were combined was measured through a normalized 90° peel test. And we used Super-high magnification lens zoom 3D microscope to observe the Interface structure around the interface and the section of interface was observed too. To profoundly study the interfacial bonding strength mechanism, using the numerical simulation method, we analysed the temperature changes of the interface. And the crystallinities in different processing parameters combination situations were monitored by thermal characterization DSC. And then we calculated the crystallinities of the melt at the interface, drawing a conclusion about the relations between crystallinity and interface strength. The results show that the process conditions are favorable for improving the crystallinity of the interface and have obvious effects on the strength of the interface.

1. Introduction

Film-insert molding (FIM), a technique also known as in-mold decoration, is a relatively new technique in which molten resin is injected to fill the cavity after a preprinted film ,which is formed as the shape of the cavity is inserted into the cavity. Recently, FIM has been widely used in injected molded parts such as automotive panels, labels for mobile phone, and household appliance. Compared with traditional injection molding, FIM is one-step process without extra post-processing including screen printing ,which can reduce manufacturing time and cost. Furthermore, the product is well protected in harsh abrasive environments with film, so the product is finished with high quality ,durability and aesthetic value. Moreover, the hot resin would wet and partially melt the film, which can provide better adhesion of the film to the substrate. It is important to stick to products through the lifetime effectivity and lasting, and the challenge for sufficient adhesion is similar to other processing such as welding, coating[1], and injection overmolding[2-6].

In the past decades, there have been many experimental studies that investigated the effect of material combination and processing parameters on the interface adhesion which determines the properties of final product. Y.W. Leong[7] measured the effect of molding conditions such as barrel temperature, injection speed and holding pressure on the morphological properties between Polypropylene(PP)-Film and PP Matrix. It was found that higher barrel temperature promotes better interface adhesion. Yamaguchi, S[8] studied the effect of crystallization on interface adhesion between PP film and PP substrate. In the latter work, He studied the adhesion between Polycarbonate (PC) Film and Polycarbonate-Acrylonitrile-Butadiene-Styrene(ABS) Substrate and he found that adhesion increased with thicker film [9]. The effect of molecular orientation induced by various injection speed on the interface was measured [10]

In these studies, fundamental understandings of the interface adhesion have been explained with one or more of following mechanism: interface interlocking, molecular diffusion and entanglement, crystallization, and adsorption theory[11-14]. To obtain the desired film-substrate interface bonding, higher melt temperature is necessary to promote molecular interpenetration, and compatible combination of film and substrate material should also be guaranteed, although the effect of holding pressure and injection speed remain uncertain in terms of different material combination. The final adhesion strength is complicated being the result of a sum of the contributions mentioned above.

For semi-crystalline polymer-polymer adhesion, besides molecular entanglement, the crystallization at or near the interface region also plays an important role in the mechanical properties and attracts a lot of attention[15, 16] .Formation of crystal depends on various factors such as nucleation location, rate of crystallization, transcrystallinity, and molding conditions[17, 18], being more complicated than amorphous polymer. However, the interface region between film and substrate is still in a rudimentary stage. Previous attempts focusing on the effect of processing conditions such as injection speed, melt temperature on the interface crystallization, the understanding of decoration film and mold temperature which determine the interface crystallization is still lacked. Thus, a deep study of the decoration film is necessary. An amorphous PC/ABS resin was in previous investigation of film's effect on adhesion. In this study, semi-crystalline resin polypropylene was used. The influence of film thickness, mold temperature on the interface morphology and adhesion was explored as well as the injection speed. The crystallinities in both the film-substrate and controlled specimen (without film) were monitored by thermal characterization DSC. 90° Peeling test was performed to gauge interface adhesion strength. Thermal analysis and Simulation were performed to get better understanding of the temperature variations at resin-film interface. The relationship between crystallinity and interfacial bond strength was established to enable the control of product properties.

2. Experiment and Methods

2.1 Materials

The substrate resin employed in this study was a commercial grade polypropylene (PP 1215C) with melting point of 165℃, density of 889kg/m^3, thermal conductivity of 0.255 W/(m·K) , and specific heat of 3100 J/(kg·K). The material of insert plastic film was polypropylene. Various thickness of the insert film used in the study were 400um, 500um, and 600um. The injection mold was made by P20 steel with density of 7850 kg/m^3, thermal conductivity of 31.5 W/(m·K) , and specific heat of 501.6 J/(kg·K).

2.2 Film-insert molding

PP substrate resin was injected to a rectangular shape(size: 140mm×60mm×2mm) using the JSW-180H.The PP film was attached to one side of the mold wall prior to injection. Adhesive tape was used at the both ends to avoid the slide of film. PP resin was injected over the film. Rectangular specimens were cut form the product to perform peel and tensile test. Specimens and product dimensions are shown in Figure. 1(a). One of the PP halves was adhered to thin-layer polyimide film to obtain a nonstick region so that the film could be detached easily from the substrate during the peeling test, the peeling test is shown in Figure. 1(b). Products devoid of the film insert were also molded as control. Injection speed, barrel temperature, film thickness and mold temperature were varied to investigate the interface adhesion. Details of the injection molding conditions are listed in table 1.

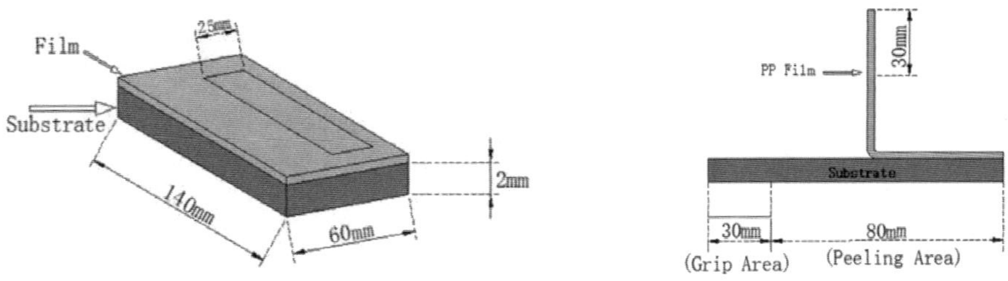

(a) (b)

Figure. 1. Injection molding species and peeling sample dimensions
(a)Injection parts; (b)peeling test pieces

Table 1. Flim injection processing conditions

Barrel temperature	220°C, 240°C, 260°C
Injection speed	10, 40, 70 mm/s
Mold temperature	30°C, 50°C, 70°C
Film thickness	0.4mm, 0.5mm, 0.6mm
Holding pressure	50 MPa
Holding time	3 s
Cooling time	60 s

2.3 Peeling and tensile test

Adhesion strength between film and substrate was gauged by a normalized 90° peel test using the zwick Proline Testing Machine. The toughness (GA) was determined by

$$G_A = \frac{P}{B}(1-\cos\theta) \tag{1}$$

where P is the peel force,B is the width of the layer and θ is the peel angle. The configuration and the dimensions of the specimens for the peel test are indicated in Fig.1b. Peeling arm was attached to the grips of a universal machine, while the fixed arm was attached to a moving plane to maintain peeling angle 90°, a peel test speed of 10 mm/min was used. This velocity ensured the test to be realized in a reasonable time period while maintaining the required alignment between specimen arms. All tests were done on at least five specimens to ensure reproducibility and accuracy of results.

2.4 Injection molding simulation

During the film-insert molding, temperature alerts the mechanical properties of final product. It is necessary to track the change of temperature field. Temperature of mold interface and melt interface can be measured by sensors[19]. However, it's difficult to track the change of polyer-film interface temperature and the polyer-film interface temperature is of vital importance to the adhesion. It's difficult to track temperature change. According to processing conditions , we simulate the effects of film on the temperature variation during film-injection molding using analysis software ANSYS and then set similar boundary and initial conditions .As the real injection process is complex, several assumptions are made: a) the initial temperature was barrel temperature, b) there are no temperature gap at the interface. In addition, we simulated the control of mold-polymer interface temperature.

2.5 Thermal and Morphological characterization

For polymer characterization of crystallization , differential scanning calorimeter (DSC) PerkinElmer DSC8000 was performed to assume thermal behavior of the materials. Heating rate of experiment was 10°C/min under the nitrogen atmosphere .We used VHX-1000C Super-high magnification lens zoom 3D microscope to observe the Interface structure around the interface and the section of interface was observed too.

3. Result and discussion

3.1 Interface temperature variation

3.1.1. Influence of mold temperature Figure 2(a) is the interface temperature change simulation in ANSYS software during the cooling period. It can be seen that at the beginning of the cooling, the temperature at the interface drops rapidly below the melting point of PP and then slowly cools down. When the temperature of the mold is 30 °C, cooling time of the interface temperature decreasing rapidly from the cylinder temperature to the melting point of PP is about 0.4s. While for mold temperature of 50°C, the time for the temperature above to decreas to melting point is about 0.5s, longer than mold temperature of 30°C. For mold temperature of 70°C, the time period is 0.9s, twice more than the time of mold temperature 30°C. When the temperature is higher then the PP melting point, PP molecules have a high degree of activity, which can carry out a lot of movement, providing a basis for better interface combination, moreover, during the cooling period, with mold temperature, resin temperatures are respectively higher leading to faster crystallization rate, which is preferred for higher crystallinity.

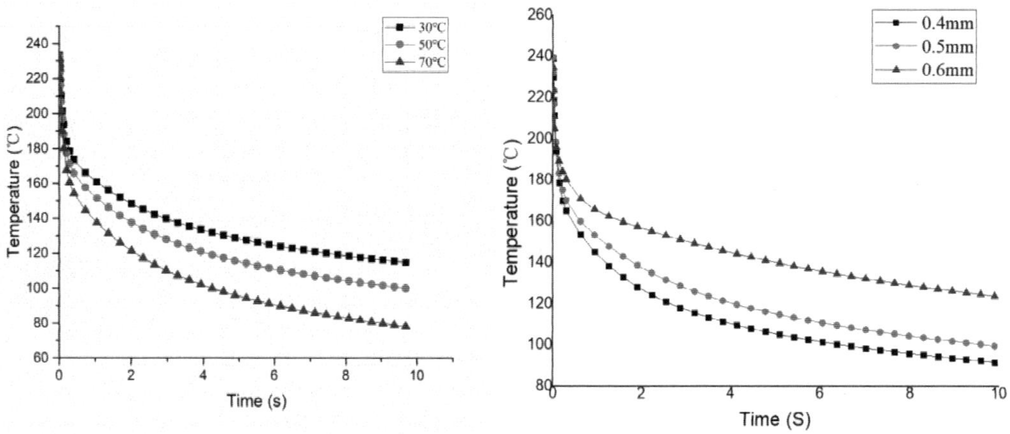

(a) (b)

Figure. 2. Interface temperature variation for different mold temperature
a)Different mold temperature; b) different film thickness

3.1.2. Influence of film thickness The presence of the film hinders heat transfer resulting in a slower temperature drop at the interface of film and PP resin. Cooling curve of the first few seconds is shown in figure 2(b).Within the first period of cooling, the temperature drop is almost the same. However with the increasing of time, the effect of film is more obvious. For thicker film, there is higher temperature at the interface. What's more, time of interface temperature above PP melting point increases from 0.37S to 0.75s responding to the film thickness from 0.4mm to 0.6mm, which is preferred for better adhesion of interface and higher crystallinity. While for interface temperature below PP melting point, molecular movement is significantly reduced, influencing little on the interface adhesion.

3.2 Interface crystallization

Injection of PP will crystallize during the cooling process.The change of the injection molding process parameters will directly affect the degree of crystallinity and grain size. Crystallization has virtual importance to adhesion strength and grain damage during peeling test absorbs more energy than amorphous structure leading to higher strength. The DSC curves of different mold temperature are shown in figure 3. Calculated crystallinity of injected species is shown in figure 8; the crystallinity of PP interface is relatively lower than core region because the melt temperature at the interface dropped quickly when coming into contact with film. There is not enough time for melting to crystallize. While for core region, the temperature is higher leading to faster crystallization speed and resulting to higher crystallinity.

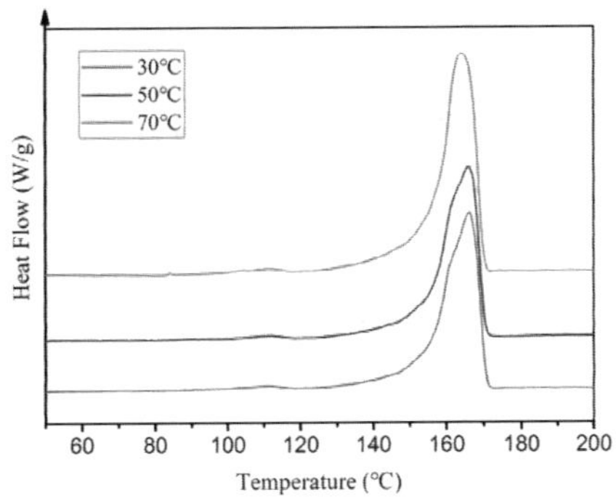

Figure. 3. DSC curves of PP at interface

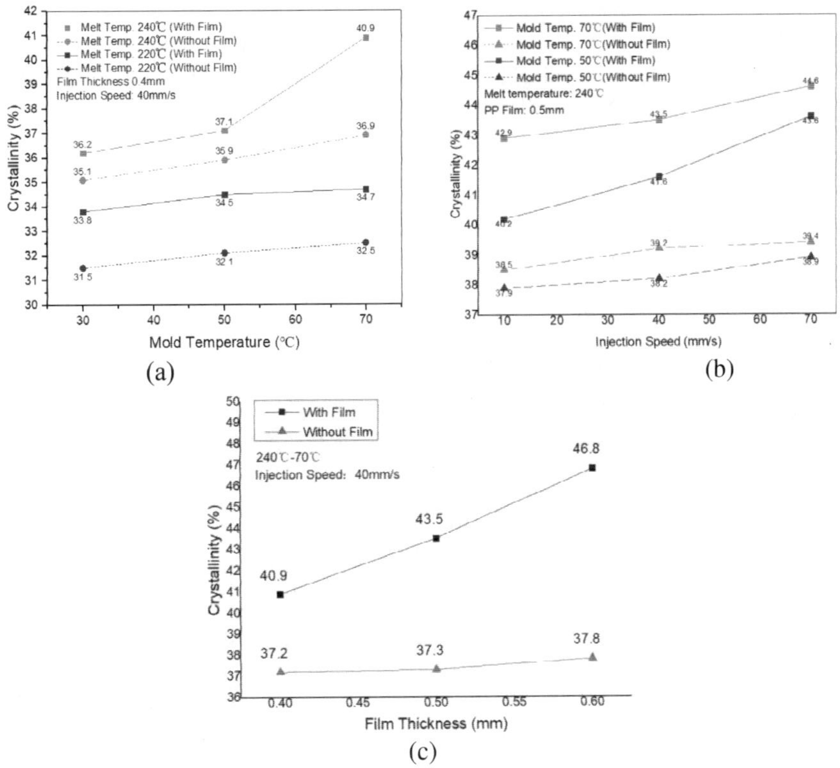

Figure. 4. Crystallinity of different injection speed and mold temperature
(a)Influence of temperature on crystallinity; (b).Influence of injection speed on crystallinity; (c).Influence of film thickness on crystallinity

The surface crystallinities of different of mold temperature are shown in figure 4(a). With mold temperature and melt temperature rising, crystallinity rises respectively, because rising of mold temperature and melt temperature accelerate crystallize rate leading to in higher crystallinity. Similar observation is shown is figure 4(b):crystallinity rises from 0.42 to 0.46 respectively with injection speed rising, for crystallization induction time is reduced and there is faster crystallize rate. Interestingly, for 240°C-50°C the crystallinity changed little while there is large fluctuation in peeling force, which probably shows that molecular movements effects more when crystallinity is relative high. The crystallinity variations of the molded parts are exhibited in Fig. 4(c) with respect to film

IOP Conf. Series: Materials Science and Engineering **436** (2018) 012023 doi:10.1088/1757-899X/436/1/012023

thickness and are compared with the parts molded without inserted film. In this figure we can observe that the crystallinity increase slightly with film thickness. Furthermore, the parts molded with film had larger crystallinity than parts molded without film. The reason can be contributed to the retardation in heat transfer that decreases polymer melt cooling rate. What's more, retardation in heat transfer is effected by film thickness. For thicker film, there is greater retardation leading to higher crystallinity which is preferred for better interface bonding, in correspondence to peeling force of film thickness showed in Fig.6.

3.3 Film–substrate adhesion

The peeling failure was of an adhesive–cohesive type showing a force–displacement peel graph like the one represented in Fig.5(c) curve. As illustrated, there are three kinds of force-displacement curve. Line (I) is relatively smooth and the force is smaller in comparison to others. It can be assumed that this type of interface bonding is of adhesive bonding which the crack generated and extended before film reached the yield stress. Interface morphology of line (I) is shown in Fig. 5(a) curve. Only a small part of film left on the substrate surface indicates the interface bonding in the preliminary stage. For line (II), a very distinct saw-teeth-like curve can be observed in cohesive type which is initiated by simultaneous crazing of the substrate and film surfaces during 90° peel test. Line (II) shows an indication of "stick–slip" phenomenon and it's believed that two factors account for the phenomenon. Firstly, the interfacial bonding is high enough so that the surface of the film would yield until a critical level, after which separation of the yielded regions will occur and the extension of crack will continue; secondly, as soon as the yield region failed, the film would "spring back" hindering crack propagation thus causing peeling force to drop suddenly, then peeling force would rise again for next crack propagation. However, if the interface adhesion is very high, film would undergo plastic deformation losing its ability to "spring back" during peeling test. The load curve of crosshead displacement illustrated as line (III) is gentle without "stick–slip" phenomenon. Morphology of the peeling interface is shown in fig. 5(b) curve, large region of the interface is covered by film, showing that bonding meets the requirements of normal use.

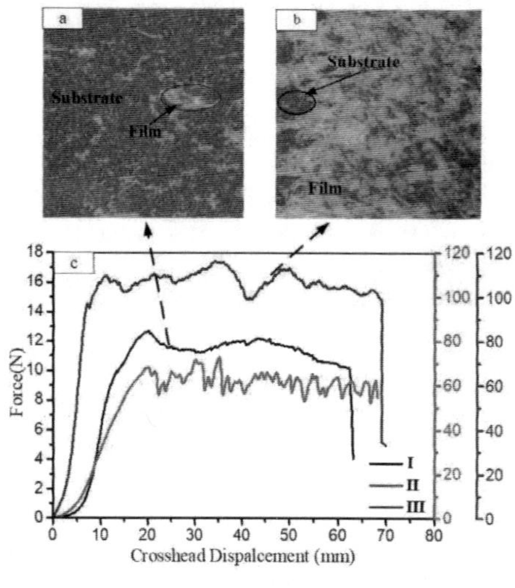

Figure. 5. Exfoliation interface ultra depth image and Peeling curve

The peeling force of different injection process parameters are showed in figure 6 and figure 7 . It shows that adhesion strength rises with melt temperature rising(showed in figure 6). For lower mold temperature the interface adhesion is relatively poor. The reason is probably that there is not efficient molecular movement at lower mold temperature. For higher mold temperature there are longer time period of temperature above PP melting point, providing active molecular entanglement leading to better interface adhesion. Figure 6. also shows that at temperature 240°C-70°C, interface adhesion is much higher than other temperature combination that achieves degree of cohesive type. Relevant peeling force of mold temperature is shown in fig.6. Compared with Fig.4, it can be assumed that adhesive bonding is effected to some extent by crystallization. A summary of the specimen conditions after the peel test is tabulated as Table 2.

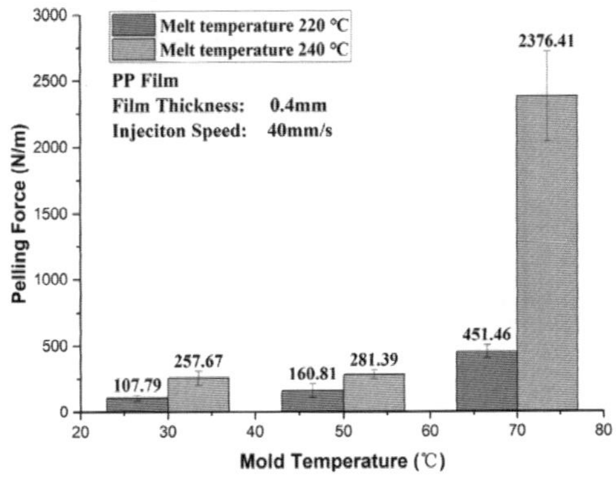

Figure. 6. Interface adhesion of different mold temperature

Table 2. Specimen Conditions after Peel Test

Melt Temp.(°C)	Injection Speed(mm/s)		
	10	40	70
220	◊	◊	◊
240	◊	◊	○

◊ adhesive bonding
○ cohesive bonding

The interfacial bonding strength at different injection speeds as shown in Figure 7. shows that, for the melt die temperature of 240 -50°C, the increase of the injection speed can promote the interfacial bonding strength. According to the results of the study of Figure 4 (b), the lifting injection speed can increase the shear effect, which is beneficial to the crystallization and the small size of the molecule. At the crystal growth, the crystallinity of the interface increases obviously, from 40.2% to 43.6%, the crystal becomes more and more energy is required to destroy the interface. In addition, the injection speed increases the temperature of the plastic melt, the film surface melts more, can also form a thicker diffusion entanglement layer, and the interfacial strength increases.

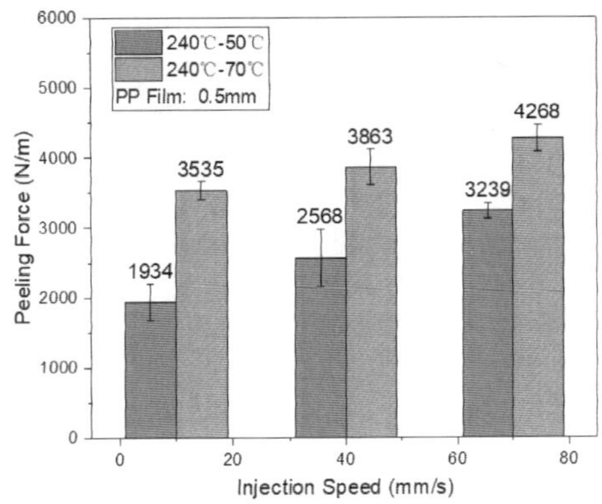

Figure. 7. Peeling force of injection speed

Koch and Mc Carren[20] stated that a thinner film was preferred to enable a firmer film–substrate adhesion in his report. Nevertheless, the findings in our work are contrary to their results and similar to the report of Leong[8]· Figure 8. shows the adhesion strength of different film thickness. It can be concluded that for thicker film there is higher interface adhesion. The existence of PP film hinders heat conduction of melting resin, causing interface temperature higher leading to blurry boundary and more active molecular movement. For higher melt and mold temperature, the interface strength is as high as ten times of lower melt and mold temperature. It seems that temperature is the primary factor affecting interface bonding strength.

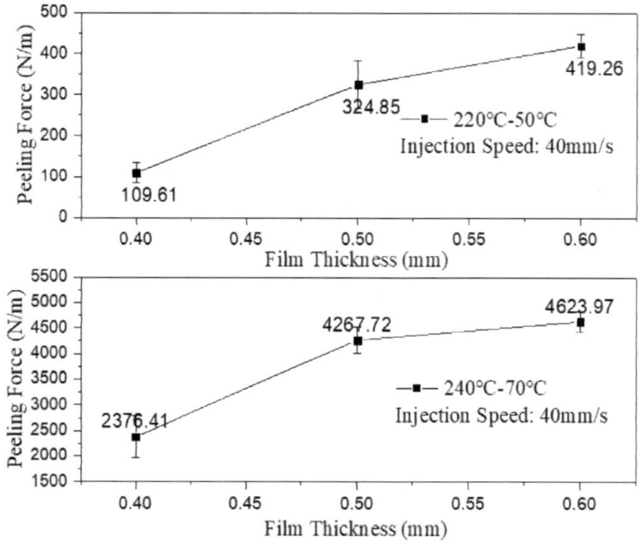

Figure. 8. Peeling force of film thickness

3.4 Interface morphology

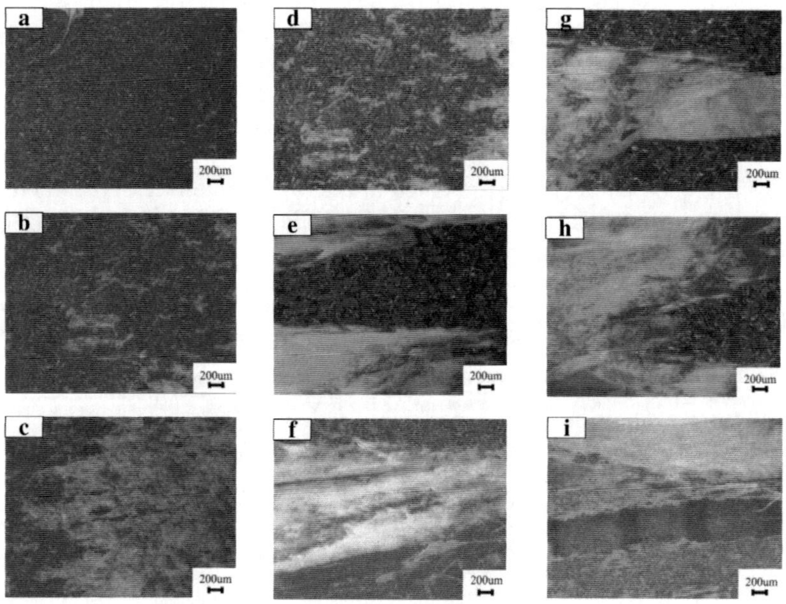

Figure. 9. Interface morphology of the substrate after 90° peeling test (a)240°C-30°C; (b) 240°C-50°C; (c) 240°C-70°C; (d) 10mm/s; (e) 40mm/s; (f) 70mm/s; (g) 0.4mm; (h) 0.5mm; (i) 0.6mm

Figure 9 shows the interface morphology of 90° peeling test under different injection conditions. Figure 9(a)-(c) shows the substrate morphology of different mold temperature with film thickness of 0.4mm. For lower temperature of 30°C there is rarely PP film left on the substrate indicating poor interface adhesion of adhesive type. But for higher temperature of 50℃, there are relatively more PP film remain after peeling test and the corresponding peeling force in figure 4 rises to 280N/mm. At mold temperature of 70℃, transition of interface bonding to cohesive type starts. Morphology indicates the mixing mode of two types. Interface morphology of different injection speed is illustrated in Figure 9(d)-(f): peeling force rise from 1729N/mm to 3739 N/mm. large area of PP film is left on the surface showing the bonding is sufficient to meet the requirements of use. In Figure 9(f) the PP film covers most area of the substrate. Figure 9(g)-9(i) is the morphology of different film thickness with melt and mold temperature of 240℃-70℃. Similar discovery is found that peeling force (shown in figure 6) is in positive correlation with cover area of PP film. The larger covered area is, the higher peeling force becomes. The reason is that for cohesive bonding the interface is melt and there are more molecular entanglement and crystallization near the interface leading to higher energy that is needed to break entanglement and crystal.

4. Conclusion

Bi-layer parts composed by PP film/substrate combination have been successfully produced by film insert molding technique. This research performs a CAE simulation investigating the effect of the film used in the IMR process on the temperature field of the mold and its impact on crystallinity as well as mold temprature. Simulation shows that for thicker film and higher mold temperature , there is higher interface temperature due to heat conduction obstruction and lower cooling rate. Results of the study show that interface bonding is significantly affected by melt temperature, injection speed, mold temperature and film thickness. Three types of peeling curves are obtained suggesting different bonding types, namely, adhesive, cohesive and mixed type. it is shown that interface crystallinity of FIM parts is altered by changing film thickness because film act as an insulator hindering heat conduction from the resin to mold, which leads to lower cooling rate. Effects of mold temperature on crystallinity are more significant due to higher mold temperature and lower cooling rate resulting in higher crystallinity. The interface crystallinity rises linearly with mold temperature rising. With the intrinsic characteristic of viscous fluids, resin temperature rises during injection process leading to faster crystallize rate. For faster injection speed, there is higher crystallinity. Our study shows that crystalliniy at the film-substrate interface region influences film-substrate adhesion. For adhesive and transition type, effects of crystallinity is significant. Higher crystallinity of the substrate due to obstructed cooling during injection molding is believed to contribute to interface bonding. Interestingly for interface bonding of cohesive type, the crystallinity rises little with injection molding parameters and molecular entanglement affects interface strength a lot due to more active molecular movement and more effective entanglement. This suggested that crystal plays a decisive role in the initial stage namely adhesive type of interface bonding and lead bonding type conversion. When it comes to cohesive boding, molecular entanglements takes the leading role dominating interface bonding strength.

References

[1] Yan, F.Y., et al., Peel-strength behavior of bilayer thermal-sprayed polymer coatings.[J] Journal of Applied Polymer Science, 2003. 88(1): p. 214-226.
[2] Nguyen, S., et al., Adhesion control for injection overmolding of elastomeric propylene copolymers on polypropylene. Effects of block and random microstructures.[J] International Journal of Adhesion and Adhesives, 2013. 46: p. 44-55.
[3] Dondero, M., et al., Adhesion control for injection overmolding of polypropylene with elastomeric ethylene copolymers.[J] Polymer Engineering & Science, 2009. 49(10): p. 1886-1893.
[4] Stan, F. and C. Fetecau. Experimental Investigation of the Adhesion Between Thermoplastic Polyurethane and Acrylonitrile-Butadiene-Styrene Substrate[C]. in ASME 2014 International Manufacturing Science and Engineering Conference collocated with the JSME 2014 International Conference on Materials and Processing and the 42nd North American Manufacturing Research Conference. 2014. : American Society of Mechanical Engineers.
[5] Rossa-Sierra, A., et al., Study of the interface behaviour between MABS/TPU bi-layer structures obtained through over moulding.[J] Materials & Design, 2009. 30(10): p. 3979-3988.

[6] Lee, H.S. and J.R. Park, Experimental study of injection-compression molding of film insert molded plates.[J] International Journal of Precision Engineering & Manufacturing, 2014. 15(3): p. 455-461.
[7] Leong, Y., et al., The effect of molding conditions on mechanical and morphological properties at the interface of film insert injection molded polypropylene-film/polypropylene matrix.[J] Polymer Engineering & Science, 2004. 44(12): p. 2327-2334.
[8] Yamaguchi, S., et al., Effect of crystallization and interface formation mechanism on mechanical properties of film-insert injection-molded poly(propylene) (PP) film/PP substrate.[J] Journal of Applied Polymer Science, 2005. 98(1): p. 294-301.
[9] Leong, Y., et al., Interfacial characteristics of film insert molded polycarbonate film/polycarbonate-acrylonitrile-butadiene-styrene substrate, part 1: Influence of substrate molecular weight and film thickness.[J] Polymer Engineering & Science, 2006. 46(12): p. 1674-1683.
[10] Leong, Y., M. Kotaki, and H. Hamada, Effects of the molecular orientation and crystallization on film–substrate interfacial adhesion in poly (ethylene terephthalate) film-insert moldings.[J] Journal of Applied Polymer Science, 2007. 104(4): p. 2100-2107.
[11] Baldan, A., Adhesion phenomena in bonded joints.[J] International Journal of Adhesion and Adhesives, 2012. 38: p. 95-116.
[12] Awaja, F., et al., Adhesion of polymers.[J] Progress in Polymer Science, 2009. 34(9): p. 948-968.
[13] Creton, C., et al., Adhesion and Fracture of Interfaces Between Immiscible Polymers: from the Molecular to the Continuum Scal, in Molecular Simulation Fracture Gel Theory. 2001, Springer. p. 53-136.
[14] Cole, P.J., R.F. Cook, and C.W. Macosko, Adhesion between immiscible polymers correlated with interfacial entanglements.[J] Macromolecules, 2003. 36(8): p. 2808-2815.
[15] Boucher, E., et al., Enhanced adhesion between polypropylene and polyamide-6: role of interfacial nucleation of the β-crystalline form of polypropylene.[J] Macromolecules, 1997. 30(7): p. 2102-2109.
[16] Laurens, C., et al., Role of the interfacial orientation in adhesion between semicrystalline polymers.[J] Macromolecules, 2001. 34(9): p. 2932-2936.
[17] Lo, C.-T., F.C. Laabs, and B. Narasimhan, Interfacial adhesion mechanisms in incompatible semicrystalline polymer systems.[J] Journal of Polymer Science Part B: Polymer Physics, 2004. 42(14): p. 2667-2679.
[18] Yan, B., et al., Interfacial crystalline structures in injection over-molded polypropylene and bond strength.[J] ACS applied materials & interfaces, 2010. 2(11): p. 3023-3036.
[19] Lin, Y.-H., et al., Effect of asymmetric cooling system on in-mold roller injection molded part warpage.[J] International Communications in Heat and Mass Transfer, 2015. 61: p. 111-117.
[20] Koch, P. and M. McCarren. Developing a Surface Texture with Film Insert Molding (FIM)(916)[C]. in TECHNICAL PAPERS OF THE ANNUAL TECHNICAL CONFERENCE-SOCIETY OF PLASTICS ENGINEERS INCORPORATED. 2000.

Synthesis Characterization and Fluorescence Property of Terbium(III) Complexes with N-phenylanthranilic Acid

Ying Wang [1,a], Huan Yuan[1], Dongyun Ma[2,b], Yunjie Zhang [1], Shujing Zhou[1] and Jie Zhang [1]

[1]College of Pharmacy, Jiamusi University, Jiamusi, Heilongjiang 154007, China
[2]Engineering Department, Fuxin Higher Training College, Fuxin, Liaoning 123001, China

[a]wxy9229@163.com; [b]mady666@sina.com

Abstract: Two binary and ternary complexes of Terbium with N-HPA and Bipy (N-HPA: N-phenylanthranilic acid, Bipy: 2,2′-bipyridine) were synthesized, and its structure was characterised by elemental analysis, UV and IR spectra. The effect of the second ligand on the fluorescence propertyof the complex was studied. The result proved that the fluoroscence intensity of the binary terbium N-HPA complex was rather weak, but the addition of the second ligand Bipy could effectively increase luminescence intensity of terbium ions. The ternary complex of $Tb(NPA)_3$ (Bipy) (NPA: radical of N-HPA by removal of one H atom) has good fluorescence property. $Tb(NPA)_3$ $(Bipy)_2$ is a promising fluorescent material, which widely used in optoelectronics engineering and technology.

1. Introduction

Some of rare earth organic complexes especialy composed by Eu^{3+} or Tb^{3+} ions with organic ligands such as β-diketones, carboxylic acids, organic supramolecular macrocycles are of promising luminescent materials[1-2]. These complexes are widely applied in luminescent materials[3,4], fluorescent probe detection[5,6], molecular imaging[7,8], displays[9,10], anti-counterfeit materials [11,12], bioscience[13-14] and other fields. Luminescent material, as we well know, the main component of light display material, is of supporting material in the field of optoelectronics engineering and technology. The optoelectronics engineering and technology plays an more and more important role in current information society. Therefore, it is of practical significance to synthesize highly efficient luminescent material and to study its luminescence mechanism. Since 1942 the luminescence of rare earth organic complex has been discovered by weissman[15], much research on them has been done[16,17]. One of the research directions is to study the effect of organic ligand on luminescence of the complex. Among many organic ligands, carboxylic acid, as mentioned above, has been widely focused on its chemical modification of substituent, and synergy with other ligands. Among carboxylic acid ligands, ligand N-phenylanthranilic acid has been widely studied. According to luminescent mechanism of rare earth organic complex, both ligand structure and energy level matching degree between ligand and rare earth ion are major factors influencing luminescence of the complex, morever appropriate second ligand can significantly influence the complex luminescence. With the second ligand 2,2′-bipyridine adding, the binary and ternary complexes of Tb-N-phenylanthranilic acid and Tb-N-phenylanthranilic acid and 2,2′-bipyridine were synthesized. Their luminescence properties were studied and compared, which provided some references for the further study of sensitized luminescence of Terbium N-phenylanthranilic acid complex with second ligand.

2. Experiment

2.1 Instruments and reagents
PE-2400 element analyzer (US PE company); Magna750 Fourier Transform Infrared Spectrometer(KBr tabletting, Nicolet, USA); 970CRT Fluorescence Spectrometer(Shanghai analytical instruments factory); UV-757 UV-Vis Spectrophotometer (SHIMADZU Shimadzu, Japan).

Determination of Terbium ion content by EDTA titration with xylenol orange as indicator.

Tb_4O_7(99.9%, China pharmaceutical group Shanghai chemical reagent Co. Ltd.), N-phenylanthranilic acid and triphenylphosphine oxide with analytical purity.

2.2 Preparation of TbCl3·6H2O
First Tb_4O_7 was completely dissolved in proper amount of concentrated hydrochloric acid changing into $TbCl_3$, second after evaporing excess hydrochloric acid in heating and stiring conditions, adding water, slowly evaporing some water until appropriate amount of solution left, the solution was cooled down, crystallized, filtered and dried, then the hydrated rare earth chloride $TbCl_3·6H_2O$ was obtained.

2.3 Synthesis of complex
First amount of $TbCl_3·6H_2O$ and N-phenylanthranilic acid(Mole ratio 1: 3) was dissolved in ethanol respectively, then mixed together. Second the solution pH was adjusted to 6-7 with ammonia water, after stirring for 4-5 hours under heating and refluxing, standing for some time, and some solid powder was obtained. Third after the powder was washed with water and ethanol 3 times respectively, and dried. Finally the white powder precipitation of binary Terbium-N-phenylanthranilic acid complex was obtained. So as to the tenary Terbium-N-phenylanthranilic acid-2,2′-bipyridine (Mole ratio 1:3:1).

3. Results and Discussions

3.1 Structural characterization of the complex
The infrared spectra of the complex $Tb(NPA)_3$ (Bipy) and ligands were shown in Fig.1.The main IR absorption data of ligands and complex was shown in table 1. After formation of the complex, the characteristic peaks of $\nu_{C=O}(1659cm^{-1})$ of the original free ligand N-HPA disappear while symmetric stretching vibration of carboxyl $\nu_{as}(coo^{-})$ at $1402cm^{-1}$ and antisymmetric stretching vibrationat peak of carboxyl $\nu_{as}(coo^{-})$ at $1633cm^{-1}$ appear. Furthermore the stretching vibration ($3335cm^{-1}$) of N-H from N-HPA presence both in ligand and complex, which indicates that the N atom from amino group does not bond with Tb^{3+}. The characteristic absorption peak of δ_{C-H} bond from ligand BIPY moves from 756 cm^{-1} to 614 cm^{-1}, meanwhile the telescopic vibration absorption peak of Tb-O appears near $540cm^{-1}$. The infrared data show that O atom from coo^{-1} of N-phenylanthranilic acid form coordination bond with Tb^{3+}. The element analysis result is showed in Table 2. According elemental analysis, composition of the ternary complex is $Tb(NPA)_3(BIPY)$, composition of the binary complex is $Tb(NPA)_3·H_2O$. Solubility test showed that $Tb(NPA)_3(BIPY)$ is soluble in chloroform, DMSO and DMF, while insoluble in water, ethanol, ether, cyclohexane, ethyl acetate, methanol and acetone.

Table 1. The main IR spectra date of ligands and complex

Vibration frequency	υ_{O-H}	$\upsilon_{c=o}$	υ_{scoo-}	υ_{ascoo-}	δ_{C-H}	υ_{Tb-O}
N-HPA	2500-2800	1659				
BIPY					756	
$Tb(NPA)_3(BIPY)$	-	-	1402	1633	614	540

Table 2. The element analysis result of the complex

complex	Elemental Analysis (measured/theoretical)			
	C%	H%	N%	Tb%
$Tb(NPA)_3(BIPY)$	61.75(62.03)	3.11(3.69)	6.79(7.42)	16.38(16.75)
$Tb(NPA)_3 \cdot H_2O$	57.29(57.57)	3.83(3.96)	5.35(5.16)	19.37(19.53)

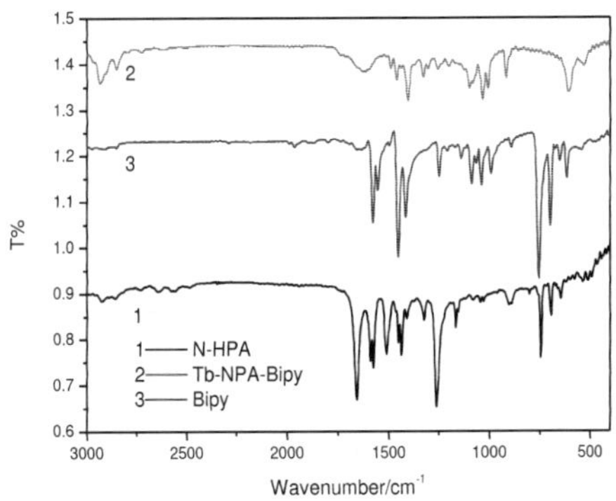

Figure.1 Infrrared spectra of the complex $Tb(NPA)_3$ (BIPY) and ligands

3.2 UV spectra of complex

The ligand and the complex were dissolved in chloroform respectively, and the UV absorption spectra were measured in range of 200-400nm.

As shown in figure 2, both the two complexes and the ligand have strong UV absorption, while the shape of the absorption peak of the complex is similar to that of the free ligand N-HPA, the relative intensity of the absorption peaks changes, and the peaks position move slightly. The absorption peaks of the ternary complex are at 288nm and 347nm, the absorption peaks of the binary complex are at 285nm and 353nm while the absorption peaks of ligand N-HPA are at 291nm and 350nm respectively. The energy absorption of the complex in ultraviolet region mainly comes from absorption of ligand N-HPA.

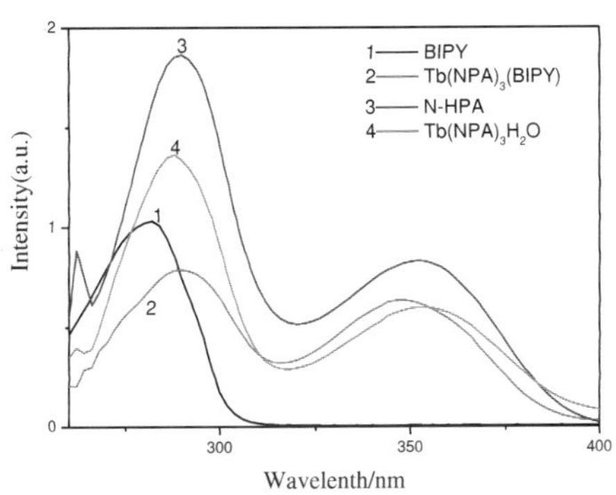

Figure.2 UV spectra of ligands and complexs

3.3 Fluorescence spectra of complex

As shown in Fig. 3, The maximum excitation wavelength of terbium binary and ternary complex is 358nm and 370nm respectively, using 545nm as monitoring wavelength. The corresponding emission spectra is obtained excited by the maximum excitation wavelength of 358nm and 370nm repectively. As shown in Fig. 4, both complexes have four emission peaks at 490nm, 546nm, 591nm and 619nm with different intensity, which are characteristic emission peaks of Tb^{3+} and attribute to $^5D_4\rightarrow{}^7F_6$,$^5D_4\rightarrow{}^7F_5$,$^5D_4\rightarrow{}^7F_4$,$^5D_4\rightarrow{}^7F_3$ transitions. Among them, the 545nm green fluorescence emission from transition $^5D_4\rightarrow{}^7F_5$ of Tb^{3+} is the strongest. Obviously stronger is the ternary complex emission intensity than the binary. Furthermore at 546nm the ternary complex emits nearly four times stronger luminenscence with narrower peak width than the binary complex, which shows that the second ligand BIPY can greatly enhance fluorescence intensity and significantly improve the fluorescence property of Terbium-N-phenylanthranilic acid complex.

According to the principle of energy transfer[16,17], there exists optimum energy matching between triplet energy level of organic ligand and the lowest excited state energy level of rare earth ion. But the energy gap between triplet energy level (24330 cm^{-1}) of the first ligand of N-HPA[18] and the lowest excited state energy level of Tb^{3+} is rather large, so the energy transfer efficiency between them is not high, morever the degree of conjugation of ligand N-HPA in Tb-NPA is poor because of steric hindrance of N atom, thus the binary complex has rather weak fluorescence. Meanwhile the lowest triplet energy level 22940cm^{-1} of the second ligand BIPY [19], matches well with energy level 20454 cm^{-1} of $^5D_4$ of Tb^{3+}, so the good energy match effectively improves energy transfer efficiency from BIPY to $Tb^{3+}$and significantly increases the fluorescence intensity of the ternary complex of $Tb(NPA)_3(BIPY)$. As a result the synthesized rare earth complex $Tb(NPA)_3(BIPY)$ with BIPY as the second ligand has excellent luminescence properties.

Another reson is that second ligand can take part or all of the place of solvent molecules involved in the coordination, therefore can form a protective layer around the rare earth complex and effectively reduce energy loss caused by the solvent molecules, prevent fluorescence quenching, therefore improve energy transfer efficiency.

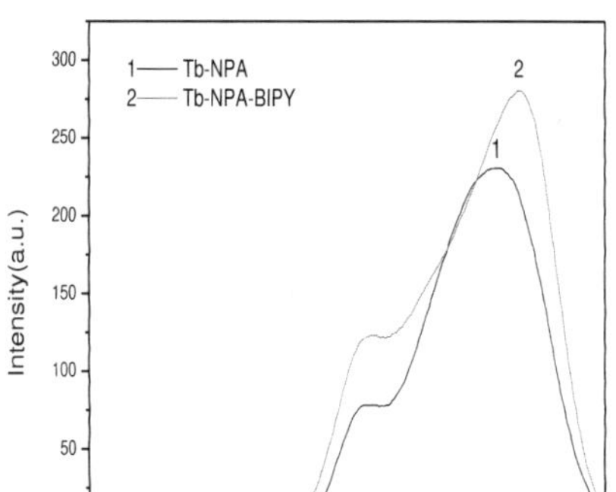

Figure.3 Exitation spectra of complexes
(1 binary complex; 2 ternary complex)

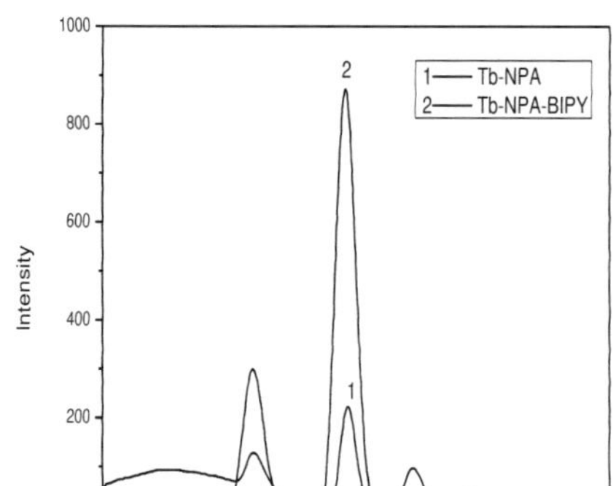

Figure.4 Fluorescence emission spectra of complexes
(1 binary complex; 2 ternary complex)

4. Conclusion

The binary complex of N-phenylanthranilic acid with Terbium and the ternary complex of N-phenylanthranilic acid and 2,2′-bipyridine with Terbium were synthesized respectively. The fluorescence intensity of the ternary complex is significant stronger than that of the binary complex, which shows that 2,2′-bipyridine as auxiliary ligand can sensitize Tb^{3+} luminescence, greatly improve the luminescence of the binary complex of Tb-N-phenylanthranilic acid. The synthesized $Tb(NPA)_3(BIPY)$ complex has good fluorescence properties. So for rare earth binary complex with weak luminescence, the addition of an appropriate second ligand can significantly increase the fluorescence intensity mutiply, resulting in excellent fluorescence of the complex.

Acknowledgements
This work was financially supported by National Natural Science Foundation of China (81601616), Heilongjiang Provincial Department of Education Basic Research Business Talent Cultivation Project (2016-KYYWF-0113), Heilongjiang Provincial Health Department Project (2016-319), Key research project of jiamusi university (12Z1201516) and Heilongjiang Provincial University Innovation and Entrepreneurship Project (201610222107).

References
[1] Y.Wang,Z.H.Jiang,Y.G.Lv,Y.J.Zhang,D.Y.Ma F.J.Zhang B.Tan, Synthetic Metals, Vol.161 (2011), p.655– 658.
[2] Y.Wang,D.C.Tong,C.X.Song,Y.L.Zhao,D.Y.Ma,Z.X.Lv,Y.G.LV, Journal of Yunnan Minzu University(Natural Sciences Edition), Vol.1 (2016) , p. 14-17
[3] D.X.Zhang,Q.Zhao,J.H.Zang,Y.J.Lu,L.Dong,C.X.Shan, Carbon Vol. 127 (2018) , p.170-176
[4] S.W. MarvadeenA,S. RochetteA,K. IshenkumbaA, A. J.Loug, Journal of Luminescence Vol.182, (2017) , p. 98–106
[5] Spectrochimica Acta Part A: Molecular and Biomolecular Spectroscopy Vol.193 (2018), p.357–364
[6] D.D.Xie,R.Y.Han,J.C.Shen,C.G.Xiao,Z.K.Zheng,Z.W.Wang, Chin J Anal Chem, Vol.43(2015), p.1069–1074
[7] L.Y.Yang, Y.P.Zhang, L.W.Hu, Y.H.Zong, R.L.Zhao, T.M.Jin, W.Gu, Journal of Rare Earths Vol. 28(2018) , p.1-6
[8] B.Tang, Z.Z.Chen, N.Zhang, J.Zhang, Y.Wang, Talanta , Vol.68 (2006) , p. 575–580
[9] J .N.Hao,B.Yan, Dalton Transactions, Vol.43,7 (2014)P.2810-2818.
[10] B Zhao,H.Zhang, Y.Q.Miao, Z.Q.Wang, L Gao, H.Wang, Y.Y.Hao, B.S.Xu, W.L.Li, J. Mater. Chem., Vol.5(2017),p.12182-12188
[11] Z.R.Li,P.Xi,M.A.Zhao,X.H.Gu,Q.Y.Ma,B.W.Cheng, Journal of rare earths, Vol.28 (2010) P.211-214
[12] L.G.Bach,B.T.P.Quynh,V.N.Thuan,X.X.Thang,K.T.Lim, J nanosci nanotechno, Vol.8, (2016)p. 8482- 8485
[13] Z.W.Deng, X.X.Hu, Y.Wwang, Y.Z.Yin, B.Peng, Z.S.Xu, J BIOMAT SCI-POLYM E, Vol.26 (2015) p.16-31
[14] X.K.Shi, , S.S Mao, K.S. Shen, H.L. Wu, X.Tang, J COORD CHEM, Vol.70 (2017), p.2015-2028
[15] Weissman S I. : The Journal of Chemical Physics, 10,4 (1942)p. 214-217.
[16] D .L.A. Dexter, The Journol of Chemical Physics, Vol.5 (1953), p.836-850.
[17] A. Sat, M. Wada,, Bulletin of The Chemical Society of Japan, Vol.43 (1970), p.1955-1962
[18] B Yan, H.J.Zhang, S.B.Wang, Chemical Journal of Chinese Universities, Vol.19(1998): p.671-675.
[19] Jason P. Cross, Peter G. Sammes, Journal of Luminescence Vol.110, 3 (2004), p. 101-111.

AUTHOR INDEX